Springer Series in Wood Science

Editor: T. E. Timell

Martin H. Zimmermann

Xylem Structure and the Ascent of Sap

With 64 Figures

Springer-Verlag
Berlin Heidelberg New York Tokyo 1983

MARTIN H. ZIMMERMANN
Harvard University
Harvard Forest
Petersham, MA 01366
U.S.A.

Series Editor:

T. E. TIMELL
State University of New York
College of Environmental
Science and Forestry
Syracuse, NY 13210
U.S.A.

Cover: Transverse section of *Pinus lambertiana* wood. Courtesy of Dr. Carl de Zeeuw, SUNY College of Environmental Science and Forestry, Syracuse, New York

ISBN 3-540-12268-0 Springer-Verlag Berlin Heidelberg New York Tokyo
ISBN 0-387-12268-0 Springer-Verlag New York Heidelberg Berlin Tokyo

Library of Congress Cataloging in Publication Data. Zimmermann, Martin Huldrych, 1926–. Xylem structure and the ascent of sap. (Springer series in wood science; v. 1) 1. Xylem. 2. Sap. 3. Plants, Motion of fluids in. I. Title. II. Series. QK871.Z55 1983 582′.01′1 83-550. ISBN 0-387-12268-0 (U.S.)

© by Springer-Verlag Berlin Heidelberg 1983
Printed in Germany.

Typesetting, printing and bookbinding: Brühlsche Universitätsdruckerei, Giessen
2131/3130-543210

This volume is dedicated to the memory of

GODFREY LOWELL CABOT

who established the Maria Moors Cabot Foundation for Botanical Research at Harvard University in 1937. He described the Foundation as follows: "The primary purpose is, by artificial selection and other methods, to increase the capacity of the Earth to produce fuel by the growth of trees and other plants. Secondly, to increase and cheapen other products of the vegetable kingdom valuable to man. Third, to disseminate information helpful and stimulating to others who may wish to enter this field of effort."

I first met Dr. Cabot in the mid-1950's when he visited my laboratory at the Harvard Forest. We were sitting on those high lab stools, and I described the work I was then doing on the translocation of sugars in the phloem of trees while he listened quietly. When I had finished he surprised me with the sudden question "How can you improve the growth of trees?" This caught me completely unprepared, because I had never thought about practical applications. After what seemed to me a rather painful silence I ventured that it would be useful to learn more about how trees function and grow. He seemed to be quite satisfied with this answer. Little did I guess that trees would be holding me under their spell for so many years!

I hope that this little book will fill the third purpose of the Cabot Foundation, namely to stimulate others who may wish to enter this field of effort.

Editor's Preface

The present volume, Xylem Structure and the Ascent of Sap by M. H. Zimmermann, very appropriately inaugurates the Springer Series in Wood Science, an enterprise recently initiated in the belief that wood and related forest products at this time have attained a new importance as renewable resources available in vast quantities. The scope of the series is intended to be wide, and virtually all aspects of wood science and technology will be considered.

Topics will include the structure of wood and bark and the chemistry of their various components, the physical and mechanical properties of wood, its formation and biodegradation, the processing of forest products, the utilization of the forest biomass, and the manufacture of pulp and paper.

Some of the volumes in this series are intended to be textbooks, but most will be monographs concerned with a limited subject area that will be treated in depth. The majority will have only one author. The books will be written by recognized experts, and will reflect the most recent information available. It is my hope that they will serve the purpose of drawing attention to wood, one of the most remarkable and useful of all natural materials.

Syracuse, New York
February, 1983

T. E. TIMELL

Acknowledgments

I am indebted to many persons who read some or all the chapters of this book: Roni Aloni, Pieter Baas, Frank Ewers, Abraham Fahn, David Fensom, Thomas McMahon, Dennis Newbanks, Barry Tomlinson, and of course the Editor of this Series, Tore Timell. The reviewers were extremely helpful in pointing out inaccuracies, drawing my attention to additional literature, etc. Discussions with Ayodeji Jeje were also very helpful. Thus, the text underwent many revisions. Although the reviews have been very helpful, I would like to stress that none of the reviewers can be made responsible for any opinions expressed in the book. Monica Mattmuller assisted me for many years in my experimental work and helped greatly in the preparation of illustrations.

M. H. ZIMMERMANN

Contents

1   Conducting Units: Tracheids and Vessels   4

1.1 Evolutionary Specialization   4
1.2 Vessel Dimensions   6
1.3 The Hagen-Poiseuille Equation and its Implications   . . . 13
1.4 Efficiency Versus Safety 15
1.5 Vessel-to-Vessel Pits 16

2   The Vessel Network in the Stem 21

2.1 Dicotyledons 21
2.2 Monocotyledons 25
2.3 A Comparison with Conifers 32
2.4 Some Quantitative Considerations 32
2.5 Implications: the Remarkable Safety Design 34

3   The Cohesion Theory of Sap Ascent 37

3.1 Negative Pressures 37
3.2 The Tensile Strength of Water 39
3.3 Tension Limits: "Designed Leaks" 44
3.4 Storage of Water 47
3.5 Sealing Concepts 54
3.6 Pressure Gradients 59
3.7 Velocities 62

4   The Hydraulic Architecture of Plants 66

4.1 The "Huber Value" 66
4.2 Leaf-Specific Conductivity 69
4.3 The Hydraulic Construction of Trees 70
4.4 Leaf Insertions 76
4.5 The Significance of Plant Segmentation 80

5   Other Functional Adaptations 83

5.1 Radial Water Movement in the Stem 83
5.2 Xylem Structure in Different Parts of the Tree 85
5.3 Wall Sculptures and Scalariform Perforation Plates 89
5.4 Aquatic Angiosperms 92

6   Failure and "Senescence" of Xylem Function 96

6.1 Embolism 96
6.2 Winter Freezing 98

6.3 Tyloses, Gums and Suberization 101
6.4 Heartwood Formation 104

7   **Pathology of the Xylem** 107

7.1 Wetwood Formation 107
7.2 Movement of Pathogens 109
7.3 The Effect of Disease on Supply and Demand of Xylem Water 113
7.4 Xylem Blockage 114
7.5 Pathogen Effects on Xylem Differentiation 123
7.6 The Problem of Injecting Liquids 123

Epilogue 127

References 129

Subject Index 141

Introduction

Wood is a marvellous tissue; it never ceases to fascinate me, be it as a construction material for buildings, ships, fine musical instruments, or as the delicate structure one sees in the microscope. Wood is described in standard plant anatomy texts (e.g., Esau 1965; Fahn 1974), and in books that are entirely devoted to it (Harlow 1970; Meylan and Butterfield 1972, 1978 a; Bosshard 1974; Core et al. 1979; Panshin and de Zeeuw 1980; Baas 1982 a, and others). Microscopic anatomy is a useful tool for wood identification, it is also of interest to biologists who are concerned with the evolution of plants, because the xylem is easily fossilized.

The study of wood function is certainly not new. Technologists have been investigating the mechanical properties of wood for years, and specific microscopic features have been recognized as serving certain aspects of water conduction. Nevertheless, the xylem as a system of pipelines, serving the plant as a whole, has received relatively little attention since Huber (1956) published his chapter on vascular water transport in the Encyclopedia of Plant Physiology. Plant anatomy and physiology have drifted apart. Plant physiology is largely concerned with biochemistry today and pays relatively little attention to the plant as a whole. Reviews of sap ascent have become so mathematical that they are not read by many plant anatomists (Pickard 1981; Hathewey and Winter 1981). Plant anatomy, on the other hand, has become largely preoccupied with evolutionary changes. But we cannot merely look at a structure and speculate what it might be good for. On the other hand, mathematical treatment should have a sound basis in structural understanding. We must study the function of structures experimentally. This is the only valid approach and it is certainly not a new one. When de Vries (1886) investigated the Casparian strip he did not just speculate about its function, he pressurized the root and tested the effectiveness of the pressure seal! We can learn a great deal from the early botanists in this respect. I was astounded how often my literature search took me back to the last century, often to ideas and approaches which have been forgotten. The reader will undoubtedly become very much aware of this fact.

The concept of evolutionary adaptation, i.e., the concept that everything about the plant has a specific purpose, can become misleading in those cases where a feature serves more than one purpose but where the investigator is preoccupied with a single one. The xylem provides a good example. In spite of this danger, the purpose of this book has been confined to xylem structure from the point of view of water conduction throughout the plant. Other functions, such as mechanical support, the periodic storage and mobilization of reserve materials in axial and radial xylem parenchyma (Sauter 1966, 1967), nutrient transport in the rays (Höll 1975) and in the axial xylem (Sauter 1976, 1980) are not covered.

I have used the old dimensions of atmosphere and bar throughout the text. The two are, for our purposes at least, practically identical (they differ by only about 1.3%). Most of my colleagues are now using the SI unit Pascal, which is the force exerted by one kg mass, at an acceleration of 1 m/sec^2 ($= 1$ Newton) per m^2 surface. I find this this unit utterly unpractical for our purposes. First, the atmosphere, 1 kg weight cm^{-2}, is so very easy to visualize. Second, it is so convenient to have ambient pressure equal one. The SI pressures are very abstract units. They may be useful for engineers, but certainly not for biologists. In the range we use them, we even have to switch back and forth between MPa and kPa, because of their awkward size. Thus I see nothing but disadvantages in using SI units in dealing with pressures in plants. I do not mind being considered either old-fashioned or reactionary for the advantage of being practical; units are supposed to be our servants, we are not their slaves! As a concession, I am giving the conversion at strategic locations of the text in parentheses. Pressures are always given as absolute values, i.e., ambient pressure is $+1$ atm (or bar), and vacuum is zero, except where specifically noted.

Writing this little book has been a very rewarding experience. As one writes one has to give the matter a great deal of thought. One often reaches suddenly a deeper level of understanding. It happened to me at least three times during writing that I suddenly thought, "ah, this is why!" First, I discovered the phenomenon of what I have subsequently called "air seeding," a mechanism by which a minute amount of air can enter a xylem duct, thus inducing cavitation (Fig. 3.6). The water-containing compartment embolizes because it has been air-seeded, but it contains only a minute amount of air. This can easily redissolve later, should the xylem pressure rise to atmospheric again. I am sure the German botanist Otto Renner, who worked on such problems in the first third of this century, was fully aware of this situation. But he never spelled it out, and I always assumed naively that air enters the cell until atmospheric pressure was reached. The second discovery was the presence of water storage in the xylem by capillarity (Chap. 3.4). I do not know if this was clear to anyone else before, but it is certainly unavoidable and must be very important. I had always known, and given lip service to the presence of intercellular spaces that are the ducts for air movement in the xylem. But it was not until we made our paint infusion experiments that I stumbled over them, so to speak, and realized that they were very conspicuous. Capillarity is an unavoidable phenomenon, and it became clear to me that one cannot have pressure changes in the xylem without changes in capillary water storage (Chap. 3.4). This led to my third major discovery. I knew that below-atmospheric pressure provides the xylem with a ready sealing mechanism (Chap. 3.5), but I had not been fully aware of the fact that below-atmospheric pressures provide the plant with water-storage space *and* air ducts in the xylem. What, then, do aquatic angiosperms do in which xylem pressures are always positive in submerged parts? Of course they must separate the two compartments! They have an air-duct system which must be sealed off from the special water-conducting "positive-pressure xylem" (Chap. 5.4). What a rich field of investigation lay ahead!

Successful competition depends on evolutionary adaptation. A few years ago Carlquist (1975) pointed out some of these problems with a book entitled *Ecological Strategies of Xylem Evolution*. This was a step in the right direction, but it also showed how very little we really know about xylem function in the whole plant. We all enjoy speculating; the reader will certainly find plenty of speculations in this volume. But what we really need is experimental evidence, and it seems to me that we have only started to scratch the surface in this respect. This book cannot be, at this stage of our ignorance, a book of encyclopedic nature. It is an "idea" book, that tries to build a bridge between the study area of structure and that of function. Functional xylem anatomy may well develop into an exciting new field of endeavor. If this is accomplished, the book will have served its purpose well.

The subject matter of this book is drawn from many different fields. It is quite obvious that I cannot be expert in all of these. Readers may therefore find errors, inaccuracies, or gaps in areas with which they are particularly familiar. I would greatly appreciate if they would communicate these to me, or let me know any comments they might have.

Chapter 1

Conducting Units: Tracheids and Vessels

1.1 Evolutionary Specialization

The development of upright land plants depended upon the development of a water-conducting system. In the beginning, water conduction and mechanical support
were closely linked, in fact this still is the case in many present-day plants that have
no vessels, like the conifers. Both water conduction and rigidity depend largely
upon cell-wall lignification, and it is thought that it was the evolution of the bio-
chemical synthesis of lignin that made upright land plants possible (Barghoorn
1964).

The first more or less continuous record of upright land plants dates back to
the upper Silurian era, about 400 million years ago (Andrews 1961; Banks 1964),
although there are sporadic earlier records. Tracheids appear to have been the only
highly specialized water-conducting elements in existence for some 300 million
years. At the end of this time, when the flowering plants arose, xylem became more
specialized by a separation of water conduction from mechanical support. This can
perhaps best be shown by the imaginative illustration of Bailey and Tupper (1918)
(Fig. 1.1). This figure suggests how, during evolution, cells became specialized as
support cells on the one hand (the fibers) and water-conducting cells on the other
(the vessel elements). In 1953 Bailey published a paper in which he discussed various aspects of the evolution of tracheary tissue.

Figure 1.1 also illustrates the significance of cell length. Tracheids in vessel-less
woods are generally not only much longer than vessel elements in vessel-containing
woods, but also longer than fibers. The fibers are serving primarily mechanical
function; this tells us that the great length attained by some tracheids before the
evolution of vessels served not a mechanical, but a hydraulic purpose. In the
flowering plants where long-distance water conduction is mostly via vessels, the
bulk of the water flows through cell series, namely the vessels, rather than through
individual cells.

Sediments of the lower Cretaceous (ca. 125 million years ago) contain fossil
wood that looks quite like modern dicotyledonous wood (Andrews 1961, p. 181;
Kramer 1974). How long such vessel-containing wood had been in existence prior
to that time is by no means certain.

The construction of a vessel is shown in Fig. 5.5, left, which illustrates three successive vessel elements of which the middle one is entire and the two outer ones are
cut open. These vessel elements (of red maple) have simple perforation plates, that
is, their end walls are completely dissolved. Figure 5.5, lower right, shows an example of a scalariform perforation plate in birch wood. End walls are not completely hydrolyzed during the final stages of development in this species. The developmental stages of vessel development, e.g., the degradation of the perforation

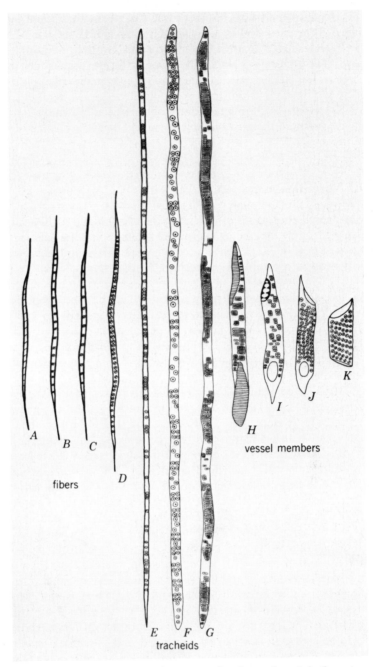

Fig. 1.1. Diagrammatic illustration of average size and structure of tracheary elements in the mature wood of some conifers and dicotyledons. **E–G** long tracheids from primitive woods (**G** showing *Trochodendron* or *Dioon*, axially foreshortened). **D–A** evolution of fibers, showing decrease in length and reduction in size of pit borders. **H–K** evolution of vessel elements, decrease in length, reduction in inclination of end walls, change from scalariform to simple perforation plates, and from scalariform to alternate vessel-to-vessel pits. (Bailey and Tupper 1918)

plate (end wall pairs) has been reviewed recently by Butterfield and Meylan (1982). The cytochemical studies of Benayoun et al. (1981) are also of interest.

Vessels show a great variety of structural features. Some are wide, others narrow, their perforation plates are of many forms, some occur in clusters, others more or less solitary, etc. This is not the place to describe this diversity in detail; there are books that illustrate it beautifully (e.g., Meylan and Butterfield 1972, 1978a). However, during the course of discussion of hydraulic properties we shall have to explore the possible functional significance of some of these features in more detail.

1.2 Vessel Dimensions

The overall dimensions of tracheids, their length and width, can be grasped relatively easily by looking at macerated xylem. Some tracheids are quite long, but those of most of our present-day conifers can at least still be shown on a single page without distortion of the proportion, by diagrammatically "folding" them (e.g., Fig. 11.6 in Esau 1965). But others are rather too long for convenient illustration. In *Agathis cunninghamii*, lengths up to 10.9 mm have been reported; a 6 mm length is even exceeded occasionally in several pine species (Bailey and Tupper 1918). In a (carboniferous) *Sphenophyllum* species, lengths range up to 3 cm (Cichan and Taylor 1982)! It would be rather difficult to illustrate these to scale.

Vessels are almost impossible to illustrate on a printed page, in fact their extent and shape was poorly known until recently. Inside vessel diameter is hydraulically an extremely important parameter; this will be discussed in the next section. Vessel diameter has been measured many times in the past. It is somewhat variable and depends, for example, upon age of tree and location within the tree (leaves, branches, trunk, etc.), a feature that will be discussed in Chapter 5.2.

It is rarely possible to see vessels throughout their entire length, because they consist of small cells which need a microscope for observation, but at the same time they are so long that a microscope is far too myopic to grasp their extent. They can never be seen on single sections or even on short series of transverse or longitudinal sections. To observe them in their entirety, we need to apply the technique of cinematographic analysis: we look at long series of wood transverse sections, recorded on film, with a special movie projector, a so-called analyzer. By running the film forward or backward, we can move up or down the stem in axial direction at any speed, go frame by frame, or stand still at any one point. The intractable axial dimension, otherwise inaccessible to the microscope, is thus translated into time, and we can move from one end of a vessel to the other end by observing a "moving" transverse stem section on the projection screen. Vessel ends thus become visible, the course of vessels can be plotted and the vessel network can be reconstructed. Cinematographic analysis will be discussed briefly in the next chapter in connection with vessel network reconstructions. Let us now look at vessel length in another way.

Vessels consist of series of individual cells, the vessel elements, whose end walls are partly or completely dissolved during late stages of cell maturation, thus forming together long capillaries. The ends usually taper out; it is very important for

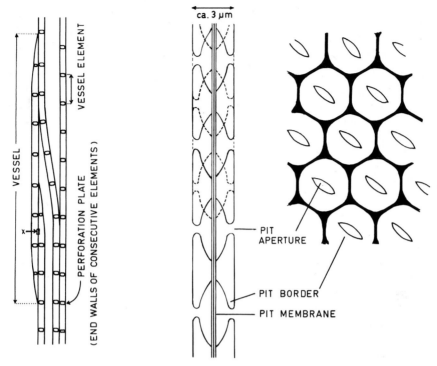

Fig. 1.2. *Left* Diagrammatic representation of a vessel network. Vessels are of finite length, their ends are overlapping. Water moves from one vessel to the next laterally through bordered pits. *Center* Diagrammatic section through a bordered pit field as we would see it at higher magnification in the *boxed area at X* of the left-hand drawing. The secondary walls arch over the primary wall pair, the pit membrane, providing mechanical strength with minimal obstruction of the membrane area. *Right* Vessel-to-vessel pit membrane in surface view. *Black hexagonal pattern* is the area where the secondary wall is attached to the primary wall, thus leaving much of the membrane area accessible to water flow. (Zimmermann and McDonough 1978)

the understanding of water conduction to realize that the water does not leave a vessel in axial direction through the very end, but *laterally* along a relatively long stretch where the two vessels, the ending and the continuing one, run side by side. The principle is shown in Fig. 1.2. A vessel consisting of nine elements is shown on the left in its entirety. Parts of three others are shown. The illustration is rather diagrammatic, because vessels consist in reality of a far greater number of elements and lengths of overlap can be very much longer. The overlap area between two vessels is of a peculiar structure, shown on the right of Fig. 1.2; this will be discussed later (Chap. 1.5).

The older literature contains only information on maximum vessel length (e.g., Greenidge 1952) and "average" vessel length (Scholander 1958). The concept of vessel-length *distribution* has been introduced by Skene and Balodis (1968). The principle of the concept is based upon the following considerations. Let us assume that the vessels of a stem are all of equal length and randomly distributed (Fig. 1.3). If we cut the stem transversely at any point, we sever individual vessels at random

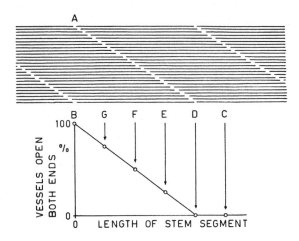

Fig. 1.3. Hypothetical stem containing vessels of equal length shown as parallel lines. Although randomly distributed in respect to their position along the stem axis, they have been rearranged in the drawing according to the location of a transverse cut made through the stem from *A* to *B*. As successive segments are cut off the right side at *C, D, E, F,* and *G,* more and more vessels are cut open at both ends. The number of these increases linearly as the stem segment is shortened

locations along their length, some near their end, some in the center, etc. Let us assume that the stem shown in Fig. 1.3 (above) has been cut across from A to B. Although randomly arranged in the stem, the vessels shown in the drawing have been rearranged according to the relation of the cut to the position of their ends, but *after* having been cut across at A–B. We now make a second cut through the stem at C. None of the vessels in the stem segment B–C will have been cut open at both ends. However, as we cut successive further disks off the right side of the stem segment at D, E, F, and G, more and more vessels will have been cut open at both ends. It is very easy to see that the number of vessels cut open at both ends increases linearly as the length of the stem segment decreases.

Let us now assume the arbitrary case where 30% of the vessels (per stem transverse section) are 18–20 cm long, and 70% are 4–6 cm long, both sizes randomly arranged. We again take a segment and cut it successively shorter until we encounter the first open vessel, this time at 20 cm stem length. The number of open 18–20-cm-long vessels increases linearly along the line A–D (Fig. 1.4) as the stem is further shortened. The first open 4–6-cm-long vessel appears at 6 cm, and their number follows the line E–F. In reality, we cannot distinguish between the vessels of different length; as we count the total number of open vessels, the count follows the line A–B–C.

This theoretical consideration can be transformed into a method for measuring vessel-length distribution. For example, one can measure the rate of air flow through a piece of wood and thereby obtain a relative measure of the number of open conduits (Zimmermann and Jeje 1981). Another method is based upon filling vessels with a suspension of particles from a cut stem end. The stem must be absolutely fresh, that is, vessels must still be water-filled when the experiment is started. This is an almost impossible requirement if the plant is under stress during a summer day, but there are various ways of overcoming the problem. One can cut a piece of stem that is much longer than needed. Air will be drawn into the vessels at both ends. Successive disks are then cut off both ends and the tension is thereby released. Another method is to vacuum-infiltrate the cut end with distilled water. A suspension of very small particles such as very dilute latex paint can then be in-

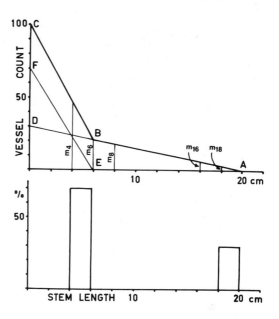

Fig. 1.4. A hypothetical wood with randomly arranged vessels of two length classes. Thirty percent of the vessels (as seen on a transverse stem section) are 18–20 cm long, 70% are 4–6 cm long. Paint has been applied *from the left* (at zero) and successive segments are cut *from the right*. A count of vessels that are cut open at both ends (indicated by containing paint), follows the line A–B–C, which is the sum of the counts A–D for the longer ones and E–F for the shorter ones. (Note that the lines E–F and B–C are not parallel). The resulting bar diagram below shows vessel-length distribution. (Zimmermann and Jeje 1981)

fused for several days. This fills all vessels to their ends with paint particles by lateral water loss (filtration), because although particles can move through a vessel and across perforation plates (Fig. 1.5, left), they are much too large to penetrate the minute pores of the pit membranes from one vessels to the next (Fig. 1.5, right).

When the infusion has been completed, the stem is cut into segments of equal length, the paint-containing vessels are counted on the cut faces of each segment and the vessel-length distribution is calculated. Let us return for this purpose to our hypothetical stem of Fig. 1.4, which was infused at its left end and divided into 2-cm-long segments. A count of the paint-containing vessels is made at various distances from the point of paint infusion, these counts are plotted and result in the line A–B–C, the sum of the counts A–D for the longer and E–F for the shorter vessels. Each count is given the designation m with the length of the stem at which they are counted as a subscript. Thus, m_{22} and m_{20} (counts at 22 and 20 cm) are zero. The first positive count is m_{18}. Beginning at the far end, we calculate the increase of the vessel count for each point. There are no vessels in the 20- to 22-cm length class. At 18 cm distance we count 3% of the total number of vessels. In order to calculate the number of vessels in the 18–20 cm length class, we carry out the following calculation:

$$[(m_{18} - m_{20}) - (m_{20} - m_{22})] \text{ times the number of steps to zero.} \tag{1.1}$$

As m_{20} and m_{22} are zero, this is equal to m_{18} times 10, i.e., 30%. The next calculation, for the 16–18 cm length class, $(m_{16} - m_{18}) - (m_{18} - m_{20})$ is zero, because the number of vessels increases linearly. The calculation continues step by step, yielding zero, until the 4–6 cm length class is reached. At this point we calculate $[(m_4 - m_6) - (m_6 - m_8)]3 = 70\%$, indicating vessels of the 4–6 cm length class. From here to zero, results are again zero. The resulting vessel-length distribution is shown in the lower half of Fig. 1.4.

100 µm

50 µm

Fig. 1.5. Longitudinal sections through metaxylem vessels of the stem of the small palm *Rhapis excelsa*, infused with a latex paint suspension. *Left* Paint particles are small enough to penetrate scalariform perforation plates. *Right* The particles are too large to penetrate the pores of the vessel-to-vessel pit membranes from the paint-containing vessel on the left to the one on the right

It is important that the reader clearly distinguishes between the two kinds of percentages referred to in this chapter, namely (1) the vessel count, i.e., the percentage of vessels that are open at a given distance from a cut across the stem (the line A–B–C in Fig. 1.4), and (2) the vessel-length distribution, the percentage of vessels in a certain length class that we calculate from the vessel count (the bar diagrams). We shall have to deal with the first when we are concerned with injuries (Chap. 2.5) and diseases (Chap. 7.2). Assume again the hypothetical stem shown in Fig. 1.4. If we make a sawcut into the tree trunk at distance zero, 15% of the vessels will be air-blocked 10 cm away from the sawcut, and 85% of the vessels will still be intact. Vessel-length distributions given in the following illustrations are all percentages of vessels in given length classes. The method of calculating percentages of vessels in length classes depends upon random distribution of vessels within the stem in which it is measured. The vessel count line should always be concave or straight, as the line A–B–C in Fig. 1.4. If it is convex, some calculated percentages are neg-

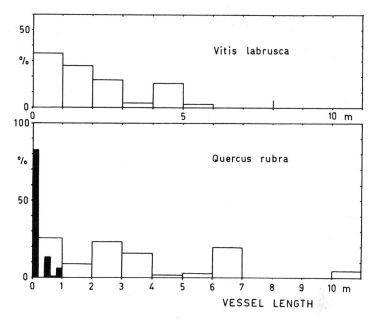

Fig. 1.6. *Above* Vessel-length distribution in the stem of the vine *Vitis labrusca. Small vertical line* at 8 m indicates longest vessel length. *Below* Vessel-length distribution in the trunk of a red oak (*Quercus rubra*). The longest vessels were in the 10–11 m length class. *Black bars* illustrate length distribution of the narrow latewood vessels. (Redrawn from Zimmermann and Jeje 1981)

ative. Negative vessel percentages may appear absurd, but can in some cases be handled. More detailed discussion will be found in the papers of Skene and Balodis (1968) and Zimmermann and Jeje (1981).

Random distribution of vessels is found in trunk wood of diffuse-porous tree species where wood tissue is relatively uniform and vessels are short. This is indicated by the fact that results are repeatable. The method works best here. In ring-porous trees some vessels may be as long as the entire trunk. Random distribution is obviously not possible in this case, and the calculation yields negative numbers, but these may be compensated for by adjacent positive numbers (Zimmermann and Jeje 1981). Other examples of non-random distributions are petioles where more shorter vessels may be present near the stem and near the blade. Branch junctions merit special attention, because in some cases vessel ends may be preferentially located at the fork. This will be discussed in more detail in connection with the concept of hydraulic architecture (Chap. 4).

Let us now look at vessel-length distribution in some species. It is quite obvious that vessel length is positively correlated with vessel diameter. This has been indicated by earlier reports on longest vessel lengths (Handley 1936; Greenidge 1952). The north temperature ring-porous trees which have large-diameter (ca. 300 μm) earlywood vessels, as well as the grapevine, whose vessel diameters are also wide, have lengths of many meters (Fig. 1.6). The vessels of diffuse-porous trees and shrubs are much narrower in diameter (of the order of 75 μm), they are also much shorter (Fig. 1.7). It is to be noted that the horizontal scales of the two illustrations

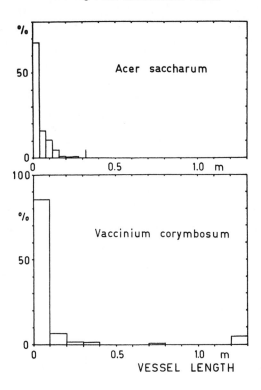

Fig. 1.7. *Above* Vessel-length distribution in the trunk of *Acer saccharum. Small vertical line* at 32 cm indicates longest length. *Below* Vessel-length distribution in the stem of the shrub *Vaccinium corymbosum.* The longest vessels were in the 1.2–1.3 m length class. (Redrawn from Zimmermann and Jeje 1981)

(Figs. 1.6 and 1.7) are quite different, and that the narrow diameter latewood vessels of oak have a length distribution not unlike that of the vessels of diffuse-porous trees. It is also quite interesting to note that the longest vessels of shrubs [several species have been investigated by Zimmermann and Jeje (1981)] are remarkably long, often as long as the shrub is tall. However, the percentages of the longest length classes are often so low that the height of the bars in the diagram does not exceed the thickness of the base line.

Figure 1.6 shows quite dramatically that the wide earlywood vessels of oak are long and the narrow latewood vessels are much shorter. In a much more subtle way this appears to be the case generally, if we can generalize from the half-dozen species that have been investigated in this respect. An *Acer rubrum*, a diffuse-porous species in which vessel diameters diminish only very slightly from early- to latewood, length also diminishes (Fig. 1.8).

Vessel-length distribution has also been reported for the small palm *Rhapis excelsa*. The vessels are comparable to those of *Acer* in width and length. We shall come back to this in a later section in connection with the vessel network.

The most important finding of this survey was the fact that xylem contains generally more short vessels than long ones. Figures 1.6 and 1.7 show this clearly. An extreme case is represented by the shrub *Ilex verticillata* in which the longest vessels were 130 cm long, but 99.5% of them were shorter than 10 cm. In another test at shorter intervals, 66% were found to be shorter than 2 cm (Zimmermann and Jeje 1981). The functional significance of the greater number of short vessels is one of safety, to be discussed in Chapter 1.4.

Fig. 1.8. Average vessel diameters (with SD) and vessel-length distributions in early- and latewood of the outermost growth ring of a 19-cm-thick trunk of *Acer rubrum*. (Zimmermann and Potter 1982)

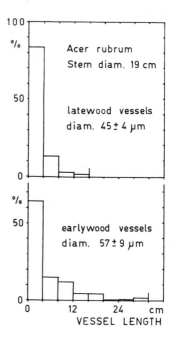

1.3 The Hagen-Poiseuille Equation and its Implications

The flow of water through vessels can be compared with flow through capillaries. We shall make this assumption, although this is not exactly right, and discuss the complications later. The flow rate (dV/dt) through a capillary is proportional to the applied pressure gradient (dP/dl) and the hydraulic conductivity (L_P),

$$\frac{dV}{dt} = L_P \frac{dP}{dl}. \tag{1.2}$$

What interests us most here is the nature of the term that describes conductivity. Conductivity is the reciprocal value of resistivity, i.e. resistance to flow, expressed per transverse-sectional area (of wood for example), but we are here really dealing with the conductance of a single capillary. Non-turbulent capillary flow was investigated independently by Hagen and by Poiseuille, who published in 1839 and 1840 respectively. Their experiments are described in some detail by Reiner (1960). Essentially, they discovered that the liquid is stationary on the capillary wall, and that its velocity increases toward the center of the tube. Visualize the following. At time zero we mark all water molecules that are located in a transverse-sectional plane in a capillary. We then let them flow and stop them again after time t. We would then find them neatly lined up on the surface of a paraboloid (Fig. 3.16). The ones touching the capillary walls have not moved at all, the ones in the center of the capillary have moved farthest. Hagen and Poiseuille found empirically that

$$L_P = \frac{r^4 \pi}{8\eta}. \tag{1.3}$$

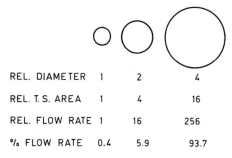

Fig. 1.9. Relative transverse-sectional areas and relative flow rates in capillaries of different diameters

REL. DIAMETER	1	2	4
REL. T. S. AREA	1	4	16
REL. FLOW RATE	1	16	256
% FLOW RATE	0.4	5.9	93.7

Parabolic flow causes the flow rate to be proportional to the fourth power of the capillary radius (r) (Reiner 1960). This is the relationship that interests us most. The other terms are constants, except for η, the viscosity of the liquid. Viscosity depends upon solute content (for example, a concentrated sugar solution is quite viscous and slows down the flow considerably). The solute concentration of xylem sap is negligible and does not measurably influence viscosity. Viscosity is also temperature-dependent, but this also need not concern us any further.

It is important to note that flow rate is proportional to the fourth power of the radius of the capillary. This means that whenever evolution brought about a slight increase in tracheary diameter, it caused a considerable increase in conductivity. Figure 1.9 shows three "vessels" in transverse section, just as they might appear in a vascular bundle observed in the microscope. Their relative diameters are 1, 2, and 4 (for example 40, 80, and 160 μm). Their relative transverse-sectional areas are 1, 4, and 16, and relative flow rates under comparable conditions, 1, 16, and 256! This tells us that if we want to compare conductivities of different woods, we should *not* compare their respective transverse-sectional vessel area, vessel density or any such measure. We must compare the sums of the fourth powers of their inside vessel diameters (or radii). If we look at vessels in xylem transverse sections, we must realize that small vessels next to large ones carry an insignificant amount of water. If the three vessels shown in Fig. 1.9 are to represent those of a vascular bundle, the smallest one would carry 0.4%, the middle one 5.9%, and the large one 93.8% of the water.

If such a slight evolutionary (and developmental) change as the widening of vessels makes them so very much more efficient, why do not all modern plants have wide vessels? The answer to this question is that efficiency comes only at a price, and in this case the price is safety. We shall look into this problem in the next section. But before we do so, we should briefly consider the question of how vessels differ from ideal capillaries.

We have seen that vessels are of finite length and that some of them are quite short. Water must move many times from one vessel to the next laterally through the membrane of the bordered pit areas. Furthermore, water must flow through perforation plates. This surely represents a considerable flow resistance if they are scalariform. Vessel walls are often not very smooth, but may contain warts, ridges and other irregularities, especially due to the presence of pits. The flow pattern is therefore not ideally paraboloid, but a good deal more complex (Jeje and Zimmermann 1979). How does all this modify flow rate?

Many years ago Ewart (1905), Berger (1931), Riedl (1937), and Münch (1943) compared the flow rates that they had experimentally measured with the flow rates that they calculated, assuming that the vessels are ideal capillaries. In some cases not flow rate but peak velocities were measured. The results were expressed in percent efficiency (see pp. 198 and 199 in Zimmermann and Brown 1971). While vines appeared to be 100% efficient (i.e., they behaved like ideal capillaries), other dicotyledonous xylem was found to be about 40% efficient (they carried about 40% as much water as ideal capillaries of the same diameter.) The result obtained with conifers was surprising. Ewart (1905) reported an efficiency of 43% for *Taxus* and Münch (1943) 26%–43% for *Abies alba*. This is quite remarkable, considering that conifers contain only tracheids. More recent investigations have been published by Tyree and Zimmermann (1971) who reported 33%–67% for red maple (*Acer rubrum*) and Petty (1978, 1981) who reported 34%–38% for birch (*Betula pubescens*), which has scalariform perforation plates, and 38% for sycamore (*Acer pseudoplatanus*).

By far the greatest problem of such measurements is the difficulty to measure inside diameter reliably. Small errors in measurement of vessel diameter yield large errors of calculated flow rates. For example a 10% error in the diameter measurement increases calculated flow rate by 46% ($1.1^4 = 1.4641$) or decreases it by 34% ($0.9^4 = 0.6561$). By the same token, we can say that xylem that has vessels which are 50% as efficient as ideal capillaries, are equal in efficiency to ideal capillaries which are only 84% as wide. In other words, it is not unreasonable, in some cases, to simply ignore the problem.

1.4 Efficiency Versus Safety

Wide vessels are very much more efficient water conductors than narrow vessels; it is therefore not surprising that evolution in many cases was toward wider vessels. However, there seems to be an upper limit of the useful vessel diameter at approximately 0.5 mm. This limit seems to have been reached many times during evolution, because there are trees and vines with wide vessels in many diverse plant families. On the other hand, the single genus *Quercus* has representatives with wide vessels (the ring-porous species of the north temperate areas) and others with narrow vessels (the evergreen species of dry or subtropical areas).

Increased vessel diameter increases efficiency of water conduction dramatically, but at the same time it decreases safety. There are various ways of looking at this. Let us compare a narrow-vessel tree like maple with a wide-vessel tree like oak, and assume that their respective vessel diameters are 75 and 300 μm. Let us assume that both trees are of equal size, transpire the same amount of water and have similarly functioning root systems. In order to carry equal amounts of water, maple must have 256 times as many vessels as oak per transverse section of the trunk. The transverse-sectional conducting area is then 16 times greater. But the vessels of oak are also much longer than those of maple. The difference can be seen in Figs. 1.6 and 1.7. As a rough estimate, we may consider the average vessel length of the wide earlywood vessels of oak to be 3 m, those of maple 10 cm. The narrow latewood vessels of oak can be ignored, because they are hydraulically insignificant as long as there are conducting earlywood vessels [as we shall see later (Chap. 6.1) this may not be the case in early spring]. Taking length into consideration, the trunk of maple must contain about $30 \times 256 = 7,680$ as many functioning vessels as that of oak. Consider now an injury, perhaps inflicted by a foraging bark beetle damaging the vessel wall at one point. Air is drawn into the injured vessel, thus blocking it permanently. The damage done in oak is 7,680 times more serious than in maple. One could argue, of course, that if the damage is limited to a specific height in the stem, maple is only 256 times as safe as oak. But in addition to this we have to consider the fact that the functioning vessels in ring-porous species are all located very near the cambium, i.e., in a superficial and therefore vulnerable position. We shall have to come back to this in the last chapter when we discuss vascular wilt diseases.

We can look at accidental damage in another way. If cavitation of the water columns were a random event, it would have to happen once in a given volume within a given time span. Maple has 16 times as much vessel volume as oak. A random event will therefore cavitate a maple vessel 16 times more often, but because of the 7,680 times greater number of vessels in maple, the damage in oak is still 480 times more serious.

Other risks that become more serious in wide-vessel woods will be discussed later. They include the possible negative correlation of tensile strength of water with compartment size (Chap. 3.3), ice formation in the winter (Chap. 6.2), and prevention of the coalescence of bubbles upon thawing of frozen xylem water (Chap. 5.3).

1.5 Vessel-to-Vessel Pits

The bordered intervessel pits are fascinating structures which we shall have to discuss repeatedly in relation to different functions. Vessel with scalariform pits (shown in Fig. 1.5, right) are assumed to be primitive; they are developmentally similar and related to protoxylem with annular or helical wall thickening. In many organs such as in the leaves of palms, one can find a continuous transition from annular or helical protoxylem tracheids to metaxylem vessels with scalariform intervessel pitting (Zimmermann and Sperry 1983).

Intervessel pits have been briefly introduced when Fig. 1.2 was discussed; they are invariably found where vessels run parallel and in contact. Scanning electron micrographs show this quite dramatically (Fig. 1.10). Regions of intervessel pitting are the locations where water moves from one vessel to the next on its way up the stem. Their characteristic bordered structure insures mechanical strength; the overarching secondary wall exposes a large membrane area (the primary wall) for water movement, but still provides support.

During vessel differentiation, intervessel pit membranes appear to swell, lose their encrusting substances and when viewed in the transmission electron microscope, become very transparent to electrons. Pit membranes between vessels and parenchyma cells show a similar but asymmetrical development (Schmid and Machado 1968). A number of papers on this topic have appeared during the years since Schmidt and Machado's original description; a recent review has been published by Butterfield and Meylan (1982). Interpretation of the chemical events is

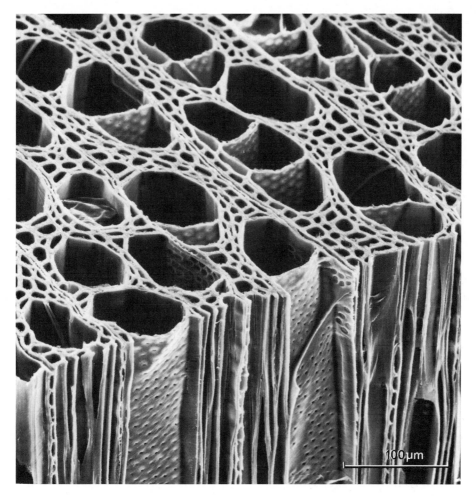

Fig. 1.10. A block of wood of *Populus grandidentata*, showing intervessel pitting on the walls between two vessels wherever they run parallel. (Scanning electron micrograph courtesy W.A. Côté)

not without controversy. Most authors believe that membrane differentiation involved hydrolysis of matrix material of the original cell wall (e.g., O'Brien and Thimann 1967), whereby a cellulosic microfibrillar web remains. However, using various histochemical techniques, a French group interprets events differently, they say that cellulose and certain hemicelluloses ("vic-glycol polysaccharides") disappear, while methylated pectins remain unchanged (Catesson et al. 1979). However the details of this process will be resolved, the important result of differentiation is increased permeability of intervessel pit membranes to water movement.

It is perhaps appropriate to mention the differentiation of coniferous bordered pits here. This has also been reviewed by Butterfield and Meylan (1982). There are of course many types, but it is sufficient if we just look at an example here (Fig. 1.11). While the torus of the pit membrane remains more or less unchanged,

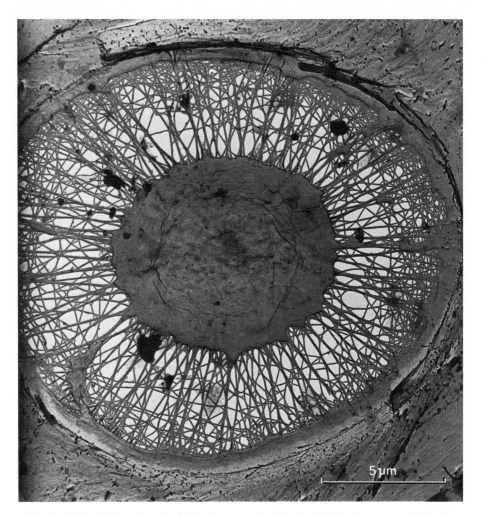

Fig. 1.11. Bordered pit of eastern hemlock (*Tsuga canadensis*), solvent-dried from green condition. The pit membrane consists of the net-like margo and the central torus. (Transmission electron micrograph courtesy W.A. Côté)

the margo loses the wall matrix substances. A cellulosic network of fibril bundles then remains which appears quite permeable. The gaps between these fibril bundles will be of particular interest to us later, because they are so wide that they not only facilitate water movement, but may let an air–water interface pass. To prevent this, the pit can act like a valve (Chap. 3.5, Fig. 3.12).

Pit areas must be relatively large in order to permit water to pass relatively freely from one vessel to the next. Their importance will be discussed in connection with the vessel network. Some tree species have what we call xylem with solitary vessels. On a transverse section, less than one out of 50 vessels can be seen to form a pair. In other words, most vessels are single and not associated with other vessels. How does the water get from one vessel to the next in this case? Are the vessels very

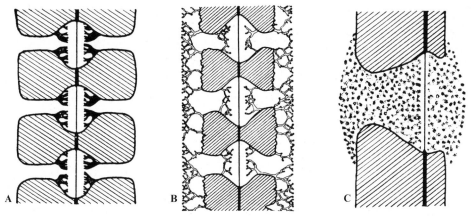

Fig. 1.12. Three drawings of sectional views of vestured pit pairs taken from Bailey (1933). **A** Coralloid outgrowth from the pit border into the pit cavity in *Combretum* sp. **B** Branched and anastomosing projections from the pit border into the pit cavity and from the inside vessel wall into the vessel lumen *(Vochysia hondurensis)*. **C** Bordered pit pair between a vessel *(left)* and a tracheid *(right)*; mats of fine texture fill the entire pit cavity *(Parashorea plicata)*

long? Are vessel overlap areas very short, thus increasing resistance to flow considerably? We do not yet know, but techniques are now available to elucidate this question.

In some species the pit borders of intervessel pits have peculiar outgrowths known as vestures. Vestured pits were decribed by Bailey (1933) in remarkable detail, considering that he had nothing more than the light microscope at his disposal. Some of Bailey's a drawings are reproduced as Fig. 1.12. In every modern textbook on wood vestured pits are illustrated with scanning electron micrographs. Detailed surveys of vestured pits were published by Ohtani and Ishida (1976) and van Vliet (1978). They show a great variety of forms and discuss their diagnostic and systematic value. Sometimes in the preparation of the specimen the pit field was split longitudinally and both split surfaces are shown, illustrating the structure dramatically (Fig. 1.13).

Intervessel pit membranes of functional vessels are not under much stress, because the pressure drop across them, even during peak flow rates, must be rather minute. However, the pressure drop across them can become very considerable as soon as one of the vessels has admitted air by injury, or has been "air-seeded" and is thus vapor-blocked (see Chap. 3.3). We then have a situation in which the pressure in the vapor-blocked vessel lumen is either atmospheric (+1 atm) or "vacuum," i.e., water vapor pressure (+0.024 atm), but in the neighboring functioning vessel is quite negative, perhaps as low as −10 or −20 atm. This enormous pressure drop across the pit membranes exerts a considerable stress on the pit membrane.

Zweypfenning (1978) suggested that the vestures are structures which support the pit membrane when they are subjected to such one-sided stress, and prevent them from tearing. Wood anatomists interested in functional adaptation might be interested to see whether vestured pits occur preferentially in trees of xeric habitat where unilateral stresses on the pit membrane can be most severe. The problem is

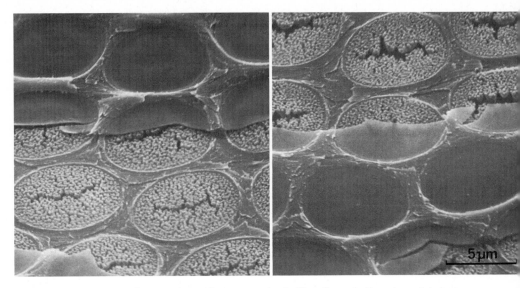

Fig. 1.13. *Lagerstroemia subcostata.* A complementary pair of split surfaces of adjacent vessel elements, showing vestured pit pairs. (Ohtani and Ishida 1976)

not quite as simple, because there are at least three additional factors which affect the danger of tearing. First, membranes of different species may differ in their mechanical properties. A membrane that is elastically or plastically deformable may be able to lodge against the pit border without tearing. Second, the pit cavity may vary in depth, the shallower it is, the sooner will the membrane come to rest on it. Third, intervessel pits vary in size considerably among species. The smaller the individual surface area, the smaller the force at a given pressure drop across it (because pressure is a force per unit area). It is thus not a simple matter to set up correlations between pit structure and habitat (Zweypfenning 1978).

Chapter 2
The Vessel Network in the Stem

2.1 Dicotyledons

The vessels of a dicotyledonous tree stem do not all run neatly parallel, they deviate more or less from their axial path. This deviation differs in successive layers of the xylem; the vessels form a network. This has been known to foresters on a macroscopic scale as "cross grain." Cross grain was considered an exceptional characteristic of certain species. We know today that virtually all xylem is microscopically "cross grained" in respect to the path of their vessels. This was first described by Braun (1959). He meticulously reconstructed the vessel network of *Populus* from series of transverse sections and showed the structure in a block of the dimensions $0.3 \times 1 \times 2.5$ mm. He also described the nature of vessel ends. Vessel ends are not easy to recognize, particularly not in single sections. They have occasionally been "seen" where India ink infusions terminated. But this is by no means reliable, because India ink particles are quite large and ragged; lateral water loss from the vessels concentrates them, and they may clog the vessels somewhere midway. Anyone who has tried to fill a fountain pen with India ink will heartily agree! Vessel ends have occasionally been reported as terminal elements in macerations (e.g., elements that have a perforation only at one end; Handley 1936; Bierhorst and Zamora 1965). This is one of the most reliable methods, but the vessel end is then seen only in isolation.

Another, even more meticulous analysis of the vessel network was reported by Burggraaf (1972). It concerns a $1.5 \times 2.0 \times 10$ mm block of European ash (*Fraxinus excelsior*). It is somewhat ironic that such an enormous effort was made at a time when a very much easier method was already available.

The introduction of cinematographic methods by Zimmermann and Tomlinson (1965, 1966) has very considerably simplified the recognition of the three-dimensional vessel structure and the localization of vessel ends. The principle is the following. A 16-mm motion picture film is assembled, frame by frame, from either serial sections in the microscope, or by photographing, with a close-up arrangement, the surface of a specimen in the microtome. The various methods are described in detail in Zimmermann (1976). Interestingly enough, the idea is about 75 years old (Reicher 1907). The important recent innovation which made the methods useful was (1) a continuous advance microtome clamp which permits the direct advance of the specimen (rather than part of the clamp), and (2) a method for aligning successive microtome sections optically in the microscope. This can be done either with an auxiliary drawing, or more elegantly with the shuttle microscope in which two adjacent sections are viewed in rapid succession. These methods transform the axial dimension of the specimen into time. With a motion picture analyzer, one can then study the three-dimensional vessel structure on a "moving" transverse section which is projected onto the table in front of the investigator. One

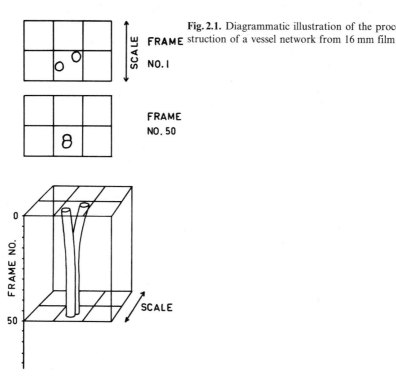

Fig. 2.1. Diagrammatic illustration of the process of reconstruction of a vessel network from 16 mm film

can move up or down in the stem by running the film forward or backward. Needless to say, such films are also very useful for teaching purposes. A film showing the three-dimensional structure of dicotyledonous wood has been available for more than 10 years (Zimmermann 1971).

The principle of vessel-network reconstruction is shown in Fig. 2.1. The film is projected onto a grid of squares; individual vessels are drawn and the frame number is recorded. The film is then slowly moved, a few frames at a time, until vessels have been displaced from their former positions, but their identity is still recognized. They are then drawn again, either in another color on the same sheet of paper, or on a new sheet. This is repeated along the desired axis length. A simple example is shown in Fig. 2.1 with film frames No. 1 and No. 50, and a grid of six squares. In a second process the vessel transverse sections are redrawn in perspective along a scale of frame numbers which represents the stem axis. The vessel network in a block of wood is thus easily reconstructed and shown in a perspective drawing. The tangential and radial scales of the block are provided by a photographed micrometer scale on the first frame of the film. The longitudinal (axial) scale is taken from the frame number, whereby each frame corresponds to the distance between individual microtome slides. In dicotyledonous wood, the axial scale is foreshortened about ten times, so that the reader has to visualize that the diagram shown should in reality be stretched about ten times to regain its natural proportions.

Figure 2.2 shows the reconstructed vessel network of cigarbox wood, *Cedrela fissilis*. Individual vessels have here been separated arbitrarily into two separate

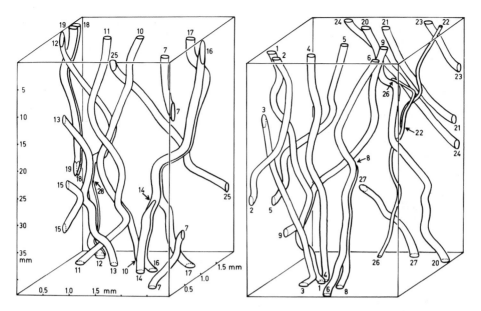

Fig. 2.2. Course of vessels in a piece of wood of *Cedrela fissilis*. Individual vessels have been arbitrarily separated into two blocks for clarity. Vessels are numbered where they exit from the block. *Arrows* show vessel ends. Note that the axial scale is foreshortened ten times. (Zimmermann and Brown 1971)

blocks so that they do not "hide" each other. In reality, all vessels were in the same block, and the block should be visualized ten times longer than it appears on paper. The vessels are numbered where they exit from the block. Numbers with arrows are vessel ends. It can be seen that all ends are in contact with other, continuing vessels. Where neighboring vessels run parallel (i.e., where they touch), we have to visualize intervessel pit areas. For example, in the center of the right-hand drawing we see the end of vessel No. 8 which runs along No. 6 for a distance of about 1.7 cm. Vessels are cut off on all faces of the block.

As one watches a vessel pair in a motion-picture film, one usually can observe them drifting apart tangentially within the growth ring. This occurs even in "straight-grained" trees like ash (*Fraxinus*). The physiological significance of this network is the tangential spreading of the axial path of water transport. This has been shown by MacDougal et al. (1929, Fig. 15) and subsequently by many others. If a dye solution is injected in a radial hole bored into a tree trunk, and the dye pattern analyzed above, one can see that it has spread tangentially within the growth ring (Fig. 2.3). This shows the axial vessel path on a much larger scale and we can see in addition to the spreading that, within each growth ring, the vessels follow a helical path up the stem. The helices usually differ from one ring to the next. This helical path is rather slight in the case shown in Fig. 2.3. The spread within individual growth rings at the application level might appear puzzling. However, we must realize that by applying the dye at ambient pressure, we introduce a very steep pressure gradient not only axially, but also laterally. Dye therefore moves up and down from vessel to vessel, penetrating the growth ring horizontally. When-

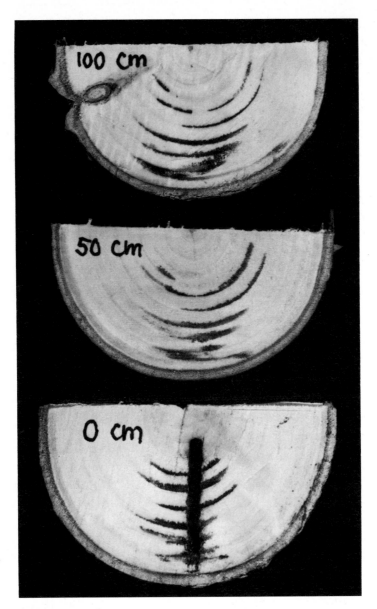

Fig. 2.3. Dye ascent from a radial hole in the trunk of *Populus grandidentata*. The dye pattern is shown in discs cut from 0, 50, and 100 cm above the point of injection. The steep pressure gradient spreads the dye tangentially even at the injection level by moving it up and down in successive vessels

ever we make a dye ascent, we alter pressure gradients and hence flow directions. For example, we always also get downward movement. Dye tracks must therefore always be interpreted with caution.

The physiological significance of this complicated layout of vessels is quite obvious: from any one root, water ascends in the trunk and spreads out within the growth rings, thus reaching not only a single branch, but a large part of the crown. Looking at it the other way, each branch of the crown obtains its water from many different roots. It is clear that this is a considerable safety feature. Loss of one or more roots does not impede the growth of individual branches, it merely diminishes the water supply to the crown slightly. It also shows that the cambium has a built-in flexibility: vessel differentiation is reoriented relatively easily and injuries can thereby by bypassed.

Spiral and wavy grain of wood have been known for a long time. However, it has only relatively recently become known that the pitch of the helical path of vessels can vary within a single growth ring. The split-disk technique shows this macroscopically. When a thin, transverse disk of a tree stem is broken along a straight line that has been scored radially on one of the transverse surfaces, the break becomes slightly wavy on the other side, because it follows the "grain" of the wood. Mariaux used this technique to recognize growth rings in tropical trees, because a very thin layer of wood at the growth ring border often follows a different helical pitch (Bormann and Berlyn 1981). This can be seen very clearly as one watches a film of *Cedrela* stem transverse sections (Zimmermann 1971). Morphogeneticists have tried to unravel the hormonal regulation of axial direction of vascular differentiation in the past (see the discussion in Burggraaf 1972). The matter has received renewed attention with the work of Polish scientists who are studying cambial domains and hormone wave patterns (Hejnowicz and Romberger 1973; Zajączkowski et al. 1983).

The exact layout of vessels in even apparently simple herbaceous dicotyledons is amazingly complex and poorly known. An entire army of plant anatomists could spend years mapping vessels, even though we now have a relatively simple method to do it. Such work would be so endless that it will have to be restricted to specific purposes such as comparative studies or for the localization of pathways for water, nutrients, or hormonal signals.

2.2 Monocotyledons

It may seem like a paradox, given the apparent complexity of the primary vasculature of monocotyledons, but the network of vessels of monocotyledons is not more difficult to study than that of dicotyledons, simply because the vessels are contained within easily identified vascular bundles. The pattern of vasculature has been analyzed in a number of arborescent monocotyledons, e.g., some palms (Zimmermann and Tomlinson 1965, 1974) *Prionium* (Juncaceae) (Zimmermann and Tomlinson 1968), *Dracaena* (Agavaceae) (Zimmermann and Tomlinson 1969, 1970), members of the Pandanaceae (Zimmermann et al. 1974), *Alpinia* (Zingiberaceae) (Bell 1980), and members of the Araceae (French and Tomlinson 1981 a–d). In spite of this relative wealth of our knowledge of the vascular structure of monocotyledonous stems, the vessel network within the vascular bundles is well known only for the small palm *Rhapis excelsa* (Zimmermann et al. 1982). However, in analyzing films of other monocotyledons, we have seen enough to recognize that *Rhapis* is representative in principle. But before we discuss the vessel network, let us first briefly review the vascular bundle network.

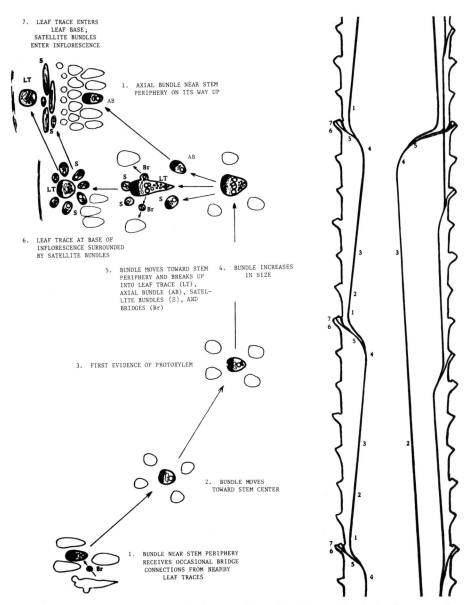

Fig. 2.4. The course of vascular bundles in the stem of the palm *Rhapis excelsa*. In the diagram on the right several bundles are shown in radial coordinate projection. The diagram is forshortened about four times in relation to the stem diameter. (Modified from Zimmermann and Tomlinson 1965)

The network of vascular bundles in the stem of *Rhapis* is less complex than in many other arborescent monocotyledons and can be regarded as the fundamental monocotyledonous pattern. However, no evolutionary meaning should be attached to the word "fundamental" here; the expression should be taken didactically. Once the vascular structure of *Rhapis* is understood, the more complex, larger palms and other monocotyledons are more easily comprehended. The following brief description is based upon the original publication of Zimmermann and Tomlinson (1965) and the more recent demonstration films (Zimmermann and Mattmuller 1982a, b).

Axial bundles run along the length of the stem. In the central part they follow a helical path which, for the sake of clarity, we shall ignore. The easiest way to illustrate the path of the bundles is in radial coordinate projection: we show them on a radial plane, after having "untwisted" the central helical part (Fig. 2.4, right). If we follow a single axial bundle up in the stem, we can see that it gradually approaches the stem center. It then quite suddenly turns outward, splits into a number of branches of which one, the leaf trace proper, enters the leaf base, another one, the continuing axial bundle, continues up the stem, and several others, the bridges, fuse with neighboring axial bundles. Additional branches connect the stem with the axillary inflorescence, but these need not concern us further here. Figure 2.4 (right) shows bundles of different lengths. There are major (axial) bundles that reach the stem center; in these the distance from one leaf contact to the next is quite long. There are minor bundles that do not reach the stem center, their leaf-contact distance is shorter. If we inspect the bundles in the microscope (as illustrated in Fig. 2.5), we see that in their lower parts they usually contain a single, relatively wide metaxylem vessel. This is shown diagrammatically in Fig. 2.4 (left) and Fig. 2.5 as Position 1 and 2. As we follow the bundle up, a point is reached where it contains a few narrow-diameter protoxylem elements (Position 3). This point is about 10 cm below the leaf contact. As we continue to follow the bundle up, the number of protoxylem elements and metaxylem vessels increase (Position 4). The bundle now turns sharply toward the stem periphery and we refer to it here as a leaf trace (this is somewhat arbitrary and is done merely for convenience). On its way toward the stem periphery, the bundle begins to break up into branches (Position 5). One (or more, or none) of these branches becomes an independent axial bundle and repeats the cycle (Position 1). Other branches are short bridges which fuse with neighboring axial bundles; still another branch, the leaf trace proper, enters the leaf base (Position 6).

What interests us particularly here is the fact that all wide metaxylem vessels remain in the stem, and that the entire complement of narrow protoxylem elements enters the leaf base. This can be seen in Fig. 2.5, and even more dramatically in a film assembled from serial sections. We can again plot individual vessels as we have done it with the dicotyledonous wood. Such an analysis, covering an entire leaf-contact distance, is illustrated in Fig. 2.6. On the right side, the radial coordinate projection is shown, on the left side the vascular bundle is projected radially onto a tangential plane. Each solid line represents a single metaxylem vessel. The hatching between individual lines indicates that the vessels run next to each other and are connected by intervessel pit areas. The dashed line indicates protoxylem, regardless of how many elements there are. The horizontal scale is very greatly expanded. If we showed the horizontal scale at the same magnification as the axial scale (which is drawn on the left and extends over a distance of 44 cm), the individual vessels would be just a fraction of a millimeter apart and the entire illustration would appear as a single line. The expanded horizontal scale lets us illustrate individual vessels and their relative location.

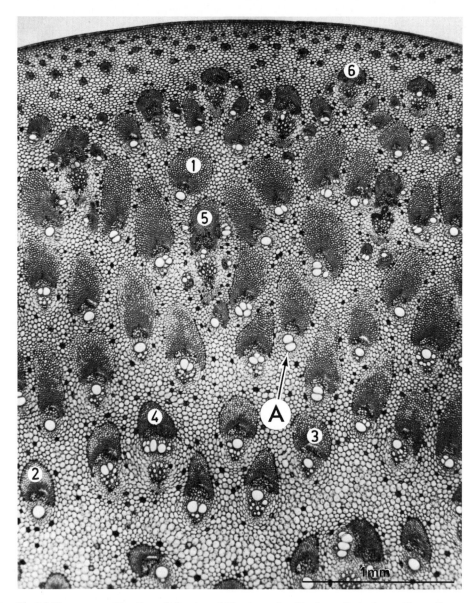

Fig. 2.5. Transverse section through the stem of *Rhapis excelsa. Numbers* refer to the positions of bundles indicated in Fig. 2.4. A vessel overlap is shown at *A*. (Modified from Zimmermann et al. 1982)

We see in Fig. 2.6 (left), for example, that the lower part of the axial bundle contains a single metaxylem vessel which is 17 cm long. At 13 and 16 cm of the scale bridges are "received" from neighboring departing leaf traces. At its lower end the 17 cm long vessel overlaps for 1 cm length with the upper end of a vessel of the leaf-trace complex. Such an overlap is seen at A in Fig. 2.5 and 2.8 (left). At its upper end it overlaps about 2.5 cm with the continuing vessel. As we study the

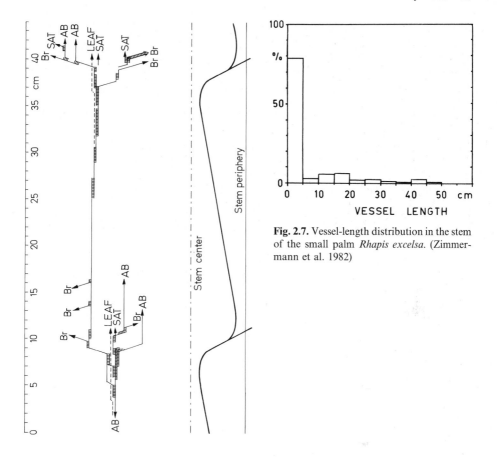

Fig. 2.7. Vessel-length distribution in the stem of the small palm *Rhapis excelsa*. (Zimmermann et al. 1982)

Fig. 2.6. Vessels of a single vascular bundle of the stem of *Rhapis excelsa*, shown over a complete leaf-contact distance of 40 cm, projected onto a tangential plane and horizontally expanded. *Solid lines* metaxylem vessels. *Hatching between solid lines* intervessel pit areas. *Dashed lines* indicate that the bundle contains protoxylem, regardless of quantity. *AB* axial bundle; *SAT* satellite bundle (leading into axillary inflorescence). *Arrows* indicating bridges (*Br*) always point away from leaf trace. (Zimmermann et al. 1982)

vessel network shown in Fig. 2.6, we realize that water that ascends in the stem has to move from one vessel to the next quite frequently. Clear vessel lengths are greater along the axial bundle, and considerably shorter in the leaf-trace complex area. On the other hand, the leaf trace-complex region provides many alternate pathways via bridges to other axial bundles.

Vessel-length distribution measured with the latex-paint method in the *Rhapis* stem show the metaxylem vessels of different parts of the vascular bundle quite nicely (Fig. 2.7). Vessels of the 0–5 cm length class are quite obviously the vessels of the leaf trace complex area (bridges, etc.), while the longer vessels are primarily those of the axial bundles in Positions 1–4.

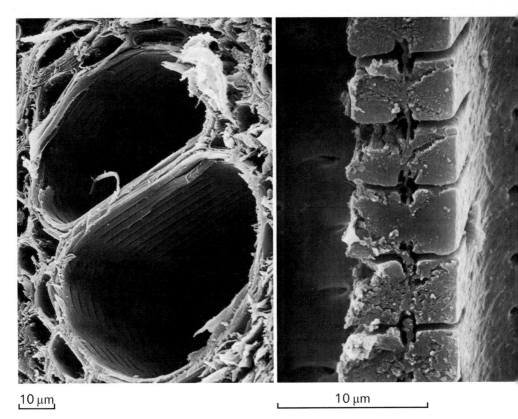

10 μm

10 μm

Fig. 2.8. Scanning electron micrographs of metaxylem vessels in the stem of *Rhapis excelsa*. *Left* transversely cut stem showing intervessel pit area. *Right* intervessel pits in longitudinal section (cutting across the wall). (Zimmermann et al. 1982)

Let us now look at the vessel-to-vessel contact in quantitative terms. Figure 2.8 (left) shows a metaxylem vessel pair in transverse section. Intervessel pits are scalariform. If we take the vessel to be circular in transverse section and its diameter to be 60 μm (Fig. 2.8, left), the transverse sectional area of the vessel lumen is 2.8×10^{-3} mm^2. The width of the scalariform pit area is ca. 35 μm. This gives us a vessel-to-vessel contact area of 0.35 mm^2 per centimeter of overlap. However, not all this area is available to water conduction. The cross section across the intervessel pit area shown in Fig. 2.8 (right) shows that ca. 40% of the contact area is pit membrane and 60% is occupied by the secondary wall. We can therefore say that 0.14 mm^2 of pit membrane area is available per centimeter of vessel overlap length for the flow of water from one vessel to the next. For a 2-cm vessel overlap the cross-sectional area through which the water must flow from one vessel to the next is ca. 100 times larger than the transverse-sectional area of a single vessel. We do not have, at this time, any information on how much this increases the resistance to flow.

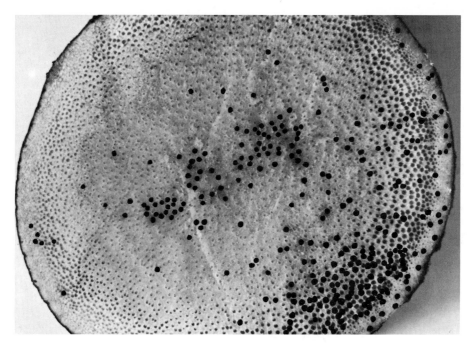

Fig. 2.9. Dye ascent from a radial hole in the stem of the palm *Chrysalidocarpus lutescens.* The disk shown was taken 25 cm above the point of injection. Stained vascular bundles are marked with a black spot. (Zimmermann and Brown 1971)

So far we have been concerned entirely with the wide metaxylem vessels and have disregarded the narrow protoxylem elements that are the pathway from stem to leaf. This is a very important point, but it will have to be considered later (Chap. 4.4).

To summarize water movement in the palm stem, we can say that it is almost exclusively via wide metaxylem vessels. In *Rhapis* and many other small palms, there is basically a single metaxylem vessel per axial bundle, such palms are called single-vessel palms. In other species, especially the larger palms, there are often two vessels ("two-vessel palms"), but these two are not connected by pit membrane areas, as there is a layer of parenchyma between them. We have not yet investigated the vessel network of any two-vessel palm (nor of any other monocotyledon!). Nevertheless, we know that water can move most freely through the long vessels of the lower part of the axial bundle. In the leaf-trace-complex area vessels are shorter, but there are several per bundle. The leaf-trace-complex area also provides direct connections, via bridges, to other bundles, and thus lateral transport paths. These are very important alternate pathways, providing the necessary by-pass pathways around injuries. A dye ascent illustrates the path of water in the stem visually. Figure 2.9 shows a disk cut from a palm stem 25 cm above a dye injection into a radial hole, similar to that illustrated in Fig. 2.3 for poplar. The dye has spread out into most of the stem area by lateral movement across bridges and by the helically twisted bundle path of the stem center.

2.3 A Comparison with Conifers

Coniferous wood consists almost entirely of tracheids. In most cases these tracheids are connected on their radial walls by bordered pit pairs. This means that water can spread easily within a growth ring in tangential direction. The axial path of the water can be traced with a dye ascent from a radial hole (Vité and Rudinsky 1959). The flow patterns made visible in this way are not too different from those of dicotyledonous trees: the dye spreads out within the growth ring and the fan-shaped dye track twists around the stem on its way up. Vité and Rudinsky (1959) found some quite distinct patterns in different groups of conifers. In the soft pine group (Haploxylon) they found what they called "left-turning spiral ascent." With this they mean that, although the dye is spreading within each growth ring, each successive growth ring has a more twisted spiral grain so that disks taken above the point of injection show a spiral track, turning clockwise on the disk from the center out. In the hard pine group (Diploxylon), larch, fir, and spruce, they found "right-turning spiral ascent" in which the spiral turns the other way. They also distinguish "interlocked," "sectorial straight," and "sectorial winding" ascent in certain other conifers. The pattern shown in Fig. 2.3 fits the last of these groups best. These classifications need not concern us further here, although it is interesting that the cambium succeeds in "tilting" successive tracheids one way or the other. The important point for us is that these patterns have the same function as the vessel network in angiosperm wood, namely that individual roots contribute to a very large part of the crown rather than to a small section of it. This is, of course, a marvellous safety feature of the xylem.

While the path of water up the tree stem is quite comparable in conifers and dicotyledons, and functionally equivalent in the monocotyledons, the lengths of the compartments, as discussed in Chapter 1, are quite different. If narrow- and short-vessel trees are less efficient but safer water conductors than the wide- and long-vessel trees, conifers are even more conservative. The mere fact that among the world's tallest trees there are conifers as well as dicotyledons shows that both designs have been successful.

2.4 Some Quantitative Considerations

It should be clear from Chapter 1.3 that a quantitative description of the xylem is only useful if the sum of the fourth powers of individual inside vessel radii (or diameters) is known. Transverse-sectional vessel area or vessel densities, etc. are of very limited value for describing water conduction capacity, though they may be useful for other purposes. In this chapter we are mainly concerned with three-dimensional aspects of structure. There are various ways in which these can be quantified.

Bosshard and Kučera (1973) give a quantitative description of the vessel network in a block $0.5 \times 1.0 \times 10$ mm of *Fagus sylvatica* taken near the cambium from a mature tree. They followed the path of 35 vessels and found a total radial spread of 0,504 mm and a total tangential spread of 0,314 mm. The first item need not concern us further at this moment because water normally remains confined within the growth ring, at least along the trunk of the tree. Movement of water across the growth ring border will be discussed later. The tangential spread of vessels of 0.314 mm over a length of 10 mm corresponds to an angular spread of ca. 1.8°. We can extrapolate and say that if the spread of 0.314 mm cm^{-1} axial length is maintained along a 10-m-tall trunk, the spread would be ca. 30 cm along the stem circumference. From dye ascents like the one illustrated in Fig. 2.3 we can also estimate the angle of tangential spread of the axial water path. It is usually of the order of 1°. It is likely that this varies quite a bit from one species to another, but it has not been systematically investigated as far as I know. Many years ago, I had the opportunity to measure the angle of tangential spread of phloem transport in *Fraxinus americana*, and found it to be of a similar magnitude, i.e., somewhat less than 1° (Zimmermann 1960).

But this is not all. As we make anatomical measurements of the possible spread of water within a growth ring, we find a fan-shaped path that, along a tall stem, would eventually enclose the entire circumference somewhere higher up in the tree. This rarely happens with a dye ascent from a point source. In Fig. 2.3 there is very little, if any spread from 50 to 100 cm height. Indeed if we crudely analyze the shape of the dye path in a stem we find the pattern to be more diamond- than fan-shaped. Although this has never been investigated in careful detail, we can suspect that the problem is one of pressure gradients. Pressure gradients will be discussed in more detail in Chapter 3.6, but we can at least say at this point that the pressures in the trunk of a transpiring tree are usually quite negative (something like −10 atm), while the dye solution we inject is at ambient pressure, i.e. +1 atm absolute. Pressure gradients in the immediate neighborhood of the injection point are therefore very steep. This means that the dye moves down as well as up, and indeed tangentially out by moving up and down repeatedly via individual vessels, thus getting away from the hole along a zigzag path. The steep pressure gradient then causes dye initially to fan out in the greatest possible angle. Far away from the injection hole, pressure gradients are not as steep and the dye stream "competes" with the water stream that comes from the roots. Very little research has been done into this direction, but it is worth giving these matters some thought, because tree injection is of considerable practical importance. We shall have to come back to it again later (Chap. 7.6).

We now have to consider the degree of mutual vessel contact: the more vessels are connected with each other, the more alternate paths for water flow are available along the stem. The longer the overlaps, the less the resistance to flow from one vessel to the next. Extreme cases are dicotyledonous wood types which show, on the one hand, many vessel clusters on a single transverse section, and on the other hand the so-called "solitary" vessel arrangement. Truly solitary vessel structure is physiologically perhaps not impossible, but would certainly be very peculiar. Research on vessel-length distribution and the three-dimensional structure of xylem with "solitary" vessels is under way (Butterfield, pers. comm.). It will be very interesting to see how water moves in stems of such trees.

Braun (1959), as well as Bosshard and Kučera (1973), made more or less complex calculations to describe the degree of vessel-to-vessel contact. Although this has given us a more precise description of wood structure than was available before, it is difficult to extract from these computations the extent of possible lateral water movement. This depends not only on the number of lateral contacts, but also upon contact lengths, which Bosshard and Kučera (1973) found to range up to ca. 2 mm in beech. The longer the contact length the better can vessel-to-vessel movement be accomplished (provided there is a pressure gradient across the pit area). Where vessels merely touch, water movement is not possible, at least not in quantity. One essential item that is missing from these analyses is vessel length. A given contact length represents a greater resistance to flow if it has to be negotiated frequently because the vessels are very short. It will be necessary to come to grips with the functional aspects of alternate pathways and resistance to flow due to vessel-to-vessel movement if we want to compare different wood species quantitatively. Undoubtedly anatomical measurements alone will not be sufficient, some experimental approaches will be necessary.

The degree to which vessels are in contact with rays has been described for beech by Kučera (1975). Vessel-to-ray contacts are the link between xylem and ray transport of water and solutes. It is interesting that Bosshard et al. (1978) found that occasional vessels ended in contact with rays rather than with other vessels. However, these vessel ends were of 10–15 µm diameter, which is rather narrow for oak, even for seedlings. Such diameters are insignificant in terms of axial flow. Nevertheless, it is developmentally interesting.

2.5 Implications: the Remarkable Safety Design

We have discussed safety of water conduction before and shall have to come back to it repeatedly. At this point we are specifically concerned with the fact that mutual vessel contact contributes very significantly to the safety of water conduction by providing alternate pathways. This is perhaps most dramatically illustrated by the so-called double sawcut experiment. If sawcuts are made along points of the stem, halfway across the tree trunk and from opposite sides, one often finds that water transport is not interrupted and that the tree survives. The concept of this experiment is very old and has been tested at least as long ago as 1806 by Cotta (cited in Hartig 1878). It has been repeated many times more recently (e.g., Preston 1952; Greenidge 1958) and is often regarded as rather mysterious and unexplicable. There is nothing mysterious about the experiment; the path of water around such injuries should by now be quite clear to the reader (see also Mackay and Weatherley 1973).

Let us first consider the problem with the sugar maple tree whose vessel-length distribution is illustrated in Fig. 1.7. If we want to know how far from a sawcut vessels are interrupted, we should not consult the vessel distribution percentages, but the vessel counts. The original vessel counts from which the distribution in Fig. 1.7 (above) was computed were the following:

Number of vessels injected at 0 cm; number of paint-containing vessels downstream

0 cm	10,786	or 100.0%	20 cm	33	0.3%
4 cm	2,147	19.9%	24 cm	12	0.1%
8 cm	762	7.1%	28 cm	2	0.02%
12 cm	229	2.1%	32 cm	0	0%
16 cm	69	0.6%			

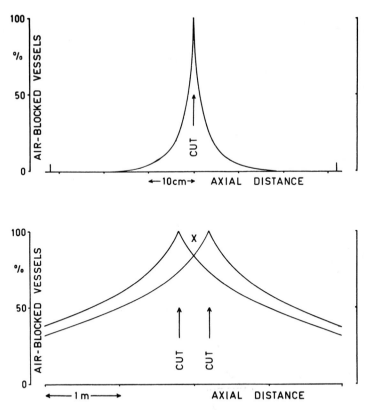

Fig. 2.10. *Above* a cut into the stem of a mature sugar maple (*Acer saccharum*) opens all vessels at that point. The graph shows the decreasing number of opened (injured) vessels at distances away from the cut. This number (the count of opened vessels) is the same as the one indicated by the line *A–B–C* in Fig. 1.4. *Small vertical bars* 32 cm from the cut indicate maximum vessel length: there are no more injured vessels beyond this point. *Below* Two sawcuts (*arrows*) made 40 cm apart into the oak (*Quercus rubra*) stem whose vessel-length distribution is illustrated in Fig. 1.7 (below). Note that the axial-distance scales of the two graphs are very different!

In other words, 20 cm from the sawcuts there are only 0.3% of the vessels air-blocked and 99.7% intact (see also Fig. 2.10 above). Consider now that the vessel path fans out from any point up or down, and also twists around the stem. This should make it clear that a double sawcut can be bypassed relatively easily. If worst comes to worst, water can always move tangentially in the tree by zigzagging up and down from vessel to vessel. We can make this visible by cutting the stem across through a radial injection hole (Fig. 2.3).

To be fair, it should be stated that some of the above-cited double sawcut experiments were made with elm, a ring-porous tree whose earlywood vessels are very long. So far, we have only one (unpublished) vessel-length-distribution measurement of a large elm tree; length distributions are comparable to those of oak or ash. Let us therefore look at the earlywood vessel counts for the oak shown in Fig. 1.6 (below). The number of vessels injected was 286. At 0.5 m distance 200 (69.9%) were containing paint, and at 1 m 158 (55.2%). Greenidge made his sawcuts 30–

40 cm apart; the interruption is rather more drastic in this case. If we assume the cuts to be 40 cm apart and plot the vessel count graphically, we can see that only about 10% of the vessels are intact 10–30 cm away from the cut (Fig. 2.10 below). The experiment would probably fail with oak, but elm has extremely wavy grain, i.e., as one follows single vessels along the stem one can see that they weave back and forth tangentially at a wavelength of some 20–50 cm. A film of transverse sections shows this very dramatically. Anyone who has tried to split elm wood knows this! The forester calls it cross grained. This means that if two sawcuts completely interrupt the trunk in axial direction, they do not interrupt all vessels because they twist around on both sides of the cuts.

Chapter 3

The Cohesion Theory of Sap Ascent

3.1 Negative Pressures

Pressures in the xylem of north temperate trees are very rarely positive. It does happen in certain species, such as *Acer*, *Betula*, *Vitis*, etc., during late winter when roots are active but leaves have not yet unfolded. If at that time the xylem is injured, it bleeds. In certain herbs, xylem pressures become positive at night when transpiration is suppressed. Leaves then guttate from hydathodes which are normally located at the leaf margins. Under humid conditions of the tropical rain forest, positive pressures have been recorded in the xylem of several species even in the presence of leaves. Guttation from leaves was then observed (e.g., von Faber 1915); walking in the forest under such conditions reportedly gave the impression of walking in a drizzle of rain.

During the growing season water is lifted up into the leaves by less than atmospheric pressures created in the leaf xylem by transpiration. In wide-vessel trees such as oaks one can hear, upon injury of the xylem by a knife or axe, the hissing sound of air being drawn into the vessels. There are many other indications that water is pulled up. Kramer (1937) showed that absorption of water by the roots lags considerably behind transpiration from leaves. Huber and Schmidt (1936) showed with their thermo-electric method that water begins to move in the most distal parts of the tree after sunrise, and later in the basal part of the trunk. In the evening when transpiration declines, movement again slows down in the branches first. When velocities in the top are plotted vs. velocities at the base of the tree, a hysteresis curve is obtained (Fig. 3.1). This happens because the tree trunk is slightly elastic and can contract, and because of the xylem's storage capacity that is provided by capillarity (Chap. 3.4). Transpiration can thus begin without instantaneous water uptake from the soil. This was also shown by Huber and Schmidt (1936), who mounted dendrographs at two different heights on tree stems, and found shrinkage first in the upper and a little later in the lower part when transpiration began in the morning. More recently such measurements have been made by Dobbs and Scott (1971) with the same results.

In doing experimental work we very often depend upon the fact that cut branches take up water when put into a container (or more poetically, when flowers are put into a vase). Moreover dye ascents almost invariably depend upon uptake of the dye solution by suction. But this does not necessarily indicate negative pressures. Indeed, we use positive absolute pressures when we pull water through a stem with a vacuum pump or when we drink a beverage through a straw. The working pressures are then between +1 atm (ambient) and near zero (vacuum), i.e., they are positive. More or less indirect methods have been used to show that xylem pressures are actually negative, i.e., below zero. In one experiment, probably first performed by Renner (1911), transpiring leaves were found to be able to pull consid-

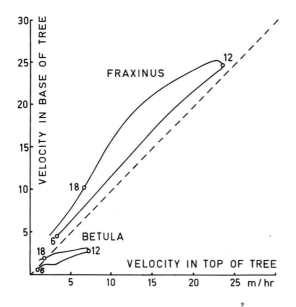

Fig. 3.1. Velocities in the upper and lower part of the tree in *Fraxinus* (a wide-vessel tree) and *Betula* (a narrow-vessel tree) during the course of 24 h. The points on the curves are 06:00 h, 12:00 h and 18:00 h. Absolute velocities are much larger in the wide-vessel species, and relative velocities are larger at the base than in the top. The reverse is true in *Betula*. However, in both species the water begins to move first in the top, and later at the bottom. (Redrawn from Huber and Schmidt 1936)

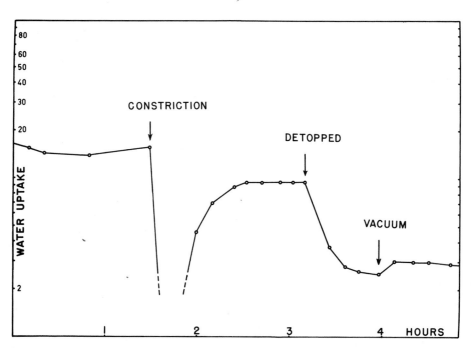

Fig. 3.2. Uptake of water by a branch from a Potometer. For 1.5 h the branch transpired freely. A constriction was then made into the axis by sawcuts and clamping. This resulted in an initial reversal of water flow at the basal end. However, transpiration continued and stabilized only at slightly lower level than before (at 2.5–3 h). The leafy part of the branch was then cut off and a vacuum pump was allowed to pull in place of it. Water consumption fell to about one third, indicating a pressure in the xylem prior to removal of the leafy part, of −2 atm. In this way Renner (1911) showed tensions up to 10 atm. (Zimmermann and Brown 1971)

erably more water through an artificially constricted stem than a vacuum pump could (Fig. 3.2). Jost (1916) showed that *Sanchezia* shoots could pull a great deal more water from their root system than a vacuum pump. It is also possible to have transpiring shoots take up water from a sealed container in which the air space above the water is evacuated (Ursprung 1913). This works only with cut shoots of species whose stem contains sufficient tracheids and short vessels to provide a path from the water-contact area to the nearest intact vessels. It usually fails with branches of ring-porous species which have wide and long vessels. Dye ascents performed under such conditions are crucial for the documentation of embolization. We shall have to discuss this in more detail in the last chapter of this book.

The cohesion theory of sap ascent is usually ascribed to H.H. Dixon, who gave a clear and detailed account in his book (Dixon 1914). But the idea that water is under negative pressures in the xylem was certainly "in the air" around 1900. Böhm (1893) probably first demonstrated water under tension in a *Thuja* twig that pulled up mercury via a water column beyond atmospheric height. For many decades after these early efforts, the cohesion theory remained controversial. Few investigators doubted that negative pressures could occur in the xylem, but many did not believe that the cohesion theory provided the full explanation of sap ascent. Two reasons caused such doubt: (1) nobody had demonstrated that negative pressures and gradients of negative pressures existed for any length of time, and (2) it was hard to visualize how water could exist in such a metastable state for long periods of time, in many cases for years. Interestingly enough, the first question was resolved by Scholander, who was one of the critics of the cohesion theory, by the introduction of the pressure bomb (Scholander et al. 1965). The pressure bomb was actually a rediscovery. Dixon (1914, pp. 142–154) had experimented with such a device. Unfortunately, as Dixon's container was made of glass and exploded twice when he used higher pressures, he discontinued the experiments. The second question, how water can remain in a metastable state for long periods of time, received little attention once negative pressures were shown to exist. The answer to this question undoubtedly involves the intricate anatomical structure of the xylem. This is described in the first two chapters of this book. Water is confined to an extremely intricate network of compartments. The fact that they are so small and so numerous makes the seemingly impossible situation possible. We also have already encountered the limit: no plant has been able to evolve vessels much in excess of 0.5 mm width. If the compartments are too large, any accidental loss becomes too serious.

3.2 The Tensile Strength of Water

It is not easy to visualize water under negative pressure. It is the same problem as visualizing water superheated above the boiling point. There is no liquid water in outer space, because it would all evaporate. When we use a vacuum pump to remove air from a container filled with water, we have no problem until the vapor pressure of water is reached. At room temperature this is 17.5 mm Hg ($=0.023$ atm or 2.3 kPa). Water begins to boil at that pressure and it is difficult or impossible to lower the pressure much further until all water has evaporated.

How can it then be subjected to zero or even to negative pressures? In an experimental set-up it is the bubbles that are trapped on the wall of the container which expand and thus expose the liquid surface from which evaporation begins. This can be observed at home: boiling in a pan on the stove can be seen to begin as steady trains of bubbles emerging always from the same points, the gas nuclei trapped on the wall of the pan. But if the liquid and container walls are entirely bubble-free, the liquid may not evaporate in the presence of negative pressure. Plants have accomplished this feat. Xylem walls are completely air-free and water uptake by roots excludes air bubbles, although not dissolved air. Water holds together quite solidly by cohesion in the myriads of small xylem compartments.

As the pressure of water drops, a point will be reached where the column finally breaks. It may do so with a click that can be made audible by proper amplification (Milburn and Johnson 1966). This is also referred to as cavitation. Cavitation is an engineering problem in the design of ship propellers. In order to propel the ship, a propeller has to "grab" water from forward and push it back. If this "grabbing" happens too vigorously, for example if the propeller runs too fast, a point may be reached where cavitation takes place and the propeller loses efficiency because it runs more or less freely in a mixture of water and vapor.

How can we show the tensile strength of water? There are indeed many ways. Let us begin with an example of the effect of air pressure. Every plant anatomist has probably had difficulties, at one time or another, to separate individual microslides or cover slips. They sometimes stick together rather tenaceously. This is not because they are actually sticky, but because the pressure of the atmosphere pushes them together. The trick then is to pry them apart at one end in order to let air (and air pressure) enter between the slides. If we do not let air in from the side, it is much more difficult to separate two pieces of glass. A normal micro slide measures 2.5×7.5 cm, its surface area is therefore 18.75 cm^2. One atmosphere (ambient air pressure) acting upon such a surface is equivalent to about 19 kg weight! In other words, without prying the slides apart at an edge to let the air pressure act between the two slides, we would have considerable difficulties in separating them.

Instead of elastic glass plates, visualize now two round steel plates with a diameter of 11.28 cm so that the surface area is exactly 100 cm^2. The plates are machined very precisely so that they are absolutely flat. We mount one of them at the ceiling of a room, flat side down, and push the other with its smooth side up to it. The plates are thick enough so that they cannot be bent easily and thus let air in from the side. A person weighing less than 100 kg (220 lbs) can hang on the lower plate without it letting go. It is held to the upper plate by air pressure. But let a second person hang on it, air pressure is now insufficient and the plate will come loose. Let us modify the experiment. We now wring the two plates together with water from which all bubbles have been removed, for example by boiling. The plates now hold together much more tenaceously by adhesive forces between steel and water, and cohesive forces within the water. In fact it is likely that an entire class of 20–50 students could be supported. An experiment similar to this was performed by Budgett (1912). He found the tensile strength of water under these conditions to be 4–60 atm, which corresponds to 400–6,000 kg weight in our 100 cm^2 plate pair described above.

Fig. 3.3. The tensile strength of water measured with a Z-shaped glass capillary of 0.6–0.8 mm diameter spinning at various velocities. *Upward pointing arrows* indicate that the water column was not broken. The remarkable result here was the ten-fold increase in tensile strength from near O° C to 5° C. (Redrawn from Briggs 1950)

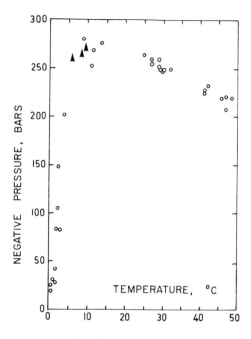

Donny (1846) was able to hang up a column of sulfuric acid in a glass tube without atmospheric support from below. This has been repeated many times with carefully prepared water columns (e.g., Dixon 1914, p. 85).

Berthelot (1850) described how he filled water into glass capillaries and sealed them with only a small vapor bubble left in the liquid. By warming up the capillary, the vapor bubble was dissolved. By cooling again, the bubble reappeared only at a distinctly lower temperature, then often with an audible click. This could be induced by a slight shock to the tube, or by rubbing. The appearing bubble expanded immediately to equilibrium size. During cooling, just before the noisy reappearance of the bubble, the water was under tensile stress. Dixon (1914) spent a considerable amount of time measuring the tensile strength of water with this method; he arrived at values of 40–200 atm. Dissolved gasses in the water did not change the results, he also did the experiments with sap that had been extracted from the xylem of plants. Many investigators repeated these experiments, obtaining somewhat different results by assuming slightly different conditions and/or modifying the calculations.

Another way to test the tensile strength is to spin water in glass capillaries in a centrifuge, thus exposing it to stress. Briggs (1950) used a Z-shaped capillary which was open at both ends. It was spun in its Z plane which was horizontal and so mounted that its center intersected the projected spin axis. Even at high speeds breakage of the water column could be observed by a marked change in refraction of the tube. The tensile strength thus found ranged from ca. 20 to 280 atm. The interesting result was the fact that these values were temperature-dependent. Between 0° and 5° C the tensile strength rose from 20 to about 260 atm; it reached a peak at ca. 10° C and then slowly declined again (Fig. 3.3). While the slow decline above 10° C is explainable, the sharp rise between 0° and 5° C is rather mysterious.

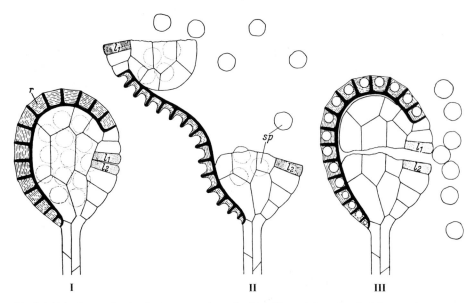

Fig. 3.4. Cohesion mechanism in the sporangium of a fern. *I* Sporangium closed; *II* Sporangium opens because transpirational loss of water from annulus cells produces tensions; *III* Increasing tension overcomes the cohesive strength of water, water vapor bubbles appear in annulus cells, the sporangium snaps closed, spores are thereby ejected. (Stocker 1952)

So far we have looked at tests of the tensile strength of water that were made in artificial containers. Inside surfaces of artificial containers have always many cracks and crevices which contain gas. Considerable efforts have been made to remove this gas by boiling or by application of high pressures before the tensile strength is tested (Apfel 1972). The stability of water in a tensile state is dependent upon the absence of bubbles of a critical size (Oertli 1971). The relative ineffectiveness of bubble removal is probably responsible for the rather erratic results of tensile strength measurements in artificial containers. We might even go a step further and say that what was measured was not the tensile strength of water, but the size to which bubbles had been reduced prior to the experiment! Calculated values of the tensile strength of water are very high [1,300–1,500 atm (Apfel 1972); 15,000 atm cited by Greenidge (1957)]. The plant evidently is able to provide much better conditions in the xylem than we can in the laboratory with glass capillaries, steel plates, and the like. The plant can grow the xylem cells walls entirely bubble free, and it takes up water via roots by the most scrupulous filtration process, even though dissolved gasses are not excluded.

Nature itself provides us with a demonstration of the tensile strength of water by the mechanism of spore ejection from fern sporangia. Certain fern sporangia possess an annulus of special cells which reaches about two thirds around the sporangium (Fig. 3.4). The walls of the annulus cells are thinner on the outside than on the inside. By the time the spores are mature, evaporation of water from the annulus cells decreases their volume, thus distorting their thin outer wall. The tension within the water of the annulus cells increases, thus forcing the sporangium open against the springy force of the annulus, until the tensile strength of the water is overcome. A vapor bubble appears in each of the cells, the sporangium snaps closed and the spores are thrown out (Fig. 3.4). Renner (1915) and Ursprung (1915) observed sporangia that they had enclosed in a small glass chamber, with a microscope. The enclosure enabled them to control the relative humidity of the surrounding air with solutions of given concentrations. By increasing the concentrations gradually the two investigators found (simultaneously and independently) that most sporangia did not snap closed until the osmotic potential of the solution had reached ca. 350 atm. In later experiments, Ziegenspeck (1928) and Haider (1954) confirmed that the water actually broke at the time of snapping; all cells contained bubbles, the first break seems to produce a jarring motion that triggers a break in all of them.

There are many aspects of tensile water that cannot be discussed here. Readers interested to learn more about this fascinating subject are referred to recent publications (Oertli 1971; Apfel 1972; Scholander 1972; Hammel and Scholander 1976). What interests us here most is, of course, how tensile strength relates to wood anatomy. It is not at all clear whether tensile strength is related to compartment size, although the largest values were obtained with fern sporangia whose annulus cells are of the order of 10–20 μm in diameter, i.e., are the smallest test compartments used by experimenters.

In 1966 Milburn and Johnson published a paper in which they described a method for the acoustic detection of vibrations produced by cavitation in xylem vessels of *Ricinus*. While suspended in air on a sound transducer, a single leaf with its petiole produced a total of roughly 3,000 clicks while it wilted. This number is not unreasonable when compared with the total number of vessels in such an organ. Click production could be stopped or slowed by adding a drop of water to the cut end of the petiole, or by covering the leaf with a polyethylene bag. It resumed when the drop or the bag was removed. Recovery (vessel refilling) was achieved by standing the petiole in water in a humid atmosphere for 24 h, but recovery was not complete, the "total click" was usually reduced by about 10% (Milburn 1973 a). Restoration was improved by vacuum infiltration (Milburn and McLaughlin 1974). Infiltration was probably into intercellular spaces, thus bringing water rapidly into contact with the embolized vessels. Vessels could hardly have been air-filled, this would have required a perforation, i.e., permanent damage. Vacuum infiltration cannot remove gas from an intact vessel whose lumen, by definition, is already at a pressure of 0.023 atm, i.e., vapor pressure of water.

What interests us at this point most is the question at what xylem pressures cavitation begins, because this is another measure for the tensile strength of water. In Milburn's experiments cavitation began at ca. 50 atm and continued as the tension was increased, via the humidity of the air by an osmotic solution, to about 160 atm. In some cases, however, cavitation began at lesser tensions. An example of this is shown in Fig. 3.10, right. These first clicks indicate relatively large pores (as we shall see later), and might actually have originated in certain intercellular spaces. But let us now consider the problem from a rather radical point of view.

The tensile strength should be a property of water, it should be great (above 1,000 atm), more or less constant, and independent of the test container. Why should there be a difference in tensile strength when measured with one (fern annulus) or the other (vascular bundles) plant material? One can easily come to the conclusion that it is impossible to measure the tensile strength of water experimentally. What we do measure is always the property of the enclosure. If we use an artificial material, such as glass capillaries, metal plates, etc., we may measure the dimensions of the gas nuclei extracted from the wall rather than the tensile strength of water. The somewhat erratic results obtained with artificial containers seem to support this idea. Briggs' (1950) results looked rather consistent, but one wonders whether his temperature dependence has also something to do with the size of gas nuclei. If, however, we use plant material for the measurement of the tensile strength of water, we may in fact measure the size of the pores in the walls of the water-containing plant cells.

3.3 Tension Limits: "Designed Leaks"

Plant cell walls might be *"designed"*, i.e., predetermined, to allow water columns to break at a certain "suitable" tension. This statement may appear unreasonable and contradict the experimental results of Ziegenspeck (1928) and Haider (1954), but I believe it does not.

Air can penetrate the pores of a wet membrane if the pores are large enough and the pressure difference on the two sides of the membrane is large enough. This is governed by the capillarity equation. The pressure required to push an air–water interface (a bubble) through a pore can be calculated from the rise of water in a vertical capillary. It must be equal to the pressure exerted by a water column of that height. The height is equal to $2T/rgs$ whereby $T =$ the surface tension of water (against air at room temperature ca. 73 dynes cm^{-1}); $r =$ the capillary radius; $s =$ the density of water; and $g =$ the acceleration of gravity. This equation can be rewritten in a simplified manner and thus made directly useful to us, namely,

$$\text{Pore Diameter [in } \mu\text{m]} \times \text{Pressure Difference [in atm]} \cong 3. \qquad (3.1)$$

This is a hyperbolic relationship which, when plotted on log-log paper, shows as a straight line (Fig. 3.5). Years ago, before the days of plastic, children who learned to swim often used a pillow case to keep themselves afloat. A pillow case will hold air rather well, provided it is thoroughly wet. Even today, one occasionally reads in emergency instructions that a wetted shirt could be used to provide floatation. The above equation shows that the smaller the pores in the fabric, the safer the floatation device, because it takes a greater pressure to squeeze out the air.

Let us now examine, step by step, what happens when the pressure difference between the outside world and a water-containing cell exceeds the sealing capacity of a wet pore (Fig. 3.6). The sealing air–water meniscus reaches a curvature with a radius that is less than the radius of the pore. It is therefore pulled through the pore toward the low pressure water-filled vessel (B). As it reaches the vessel lumen it breaks the water column as water evaporates into it explosively (C). The growing bubble volume causes the pressure inside the vessel to rise momentarily, the pore will therefore reseal (D). The bubble whose diameter is growing because of water evaporating into it (E) will rapidly fill the entire vessel (F). It is possible that the bubble does not separate from the wall as shown in D and E. However, the effect

Fig. 3.5. Graphic representation of the capillarity equation

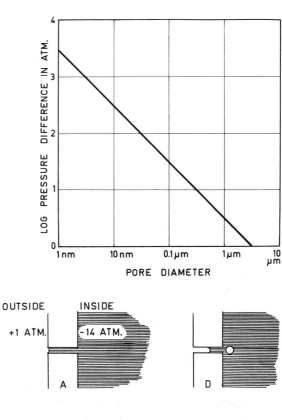

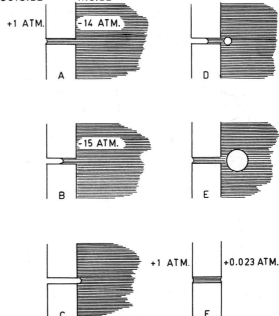

Fig. 3.6. "Air-seeding" and embolism of a vessel containing water under stress. A water-filled vessel under − 14 atm pressure (**A**). Its wall is punctured by a hole of about 0.2 μm diameter. As the pressure drops, air is admitted (**B**). As soon as the meniscus reaches the vessel lumen (**C**), water evaporates into it explosively. The bubble thereby enlarges (**D**), thus raising the pressure inside the vessel and resealing the pore. The bubble rapidly increases in size (**E**) until the entire vessel is vapor-filled (**F**). The bubble may not detach from the wall as shown in **D** and **E**; water would then move into the pore from within the wall, and close it. The final pressure in the vessel will be very near the water vapor pressure (i.e., ca. 0.023 atm), because only a very tiny amount of air, i.e., a bubble of only a fraction of a μm in diameter, has been admitted before the pore resealed when the pressure difference diminished

would be the same: the pore would refill by water influx from the wall. As only a tiny original air seed entered the vessel, i.e., a bubble a fraction of a µm in diameter, the final pressure in the vessel will approximate water vapor pressure, i.e., ca. 0.023 atm (at room temperature). It was found many years ago that not only annulus cells of fern sporangia, but similar cells of other plants that operate with cohesion mechanisms, are water vapor-, not air-filled after snapping [Ziegenspeck (1928), and the literature cited therein].

An interesting aspect of this mechanism is the fact that a water-containing cell with a perforated wall in which the pressure drops below the sealing limit given by the pore sizes will instantaneously reseal after admitting a bubble seed. This means that if the xylem pressure later rises, the cell will progressively refill (Chap. 3.4). If the pressure rises above atmospheric, it may completely refill. This has been described in the previous section as the "recovery" in Milburn's click experiments. If positive pressures occur not too long after embolism has taken place, refilling can be perfect. If the cell remains too long under less than atmospheric pressure, it is possible that dissolved gases diffuse into the space, and after the cell has been refilled with water, a residual gas bubble may remain. Above atmospheric pressures can redissolve such gas bubbles if they are not too big. We are concerned here not with actual injuries, which will be discussed in the last two chapters, but with cavitation due to pore sizes which are part of the plant's "design." In other words we consider here the possibility of "designed" leaks in plant cell walls.

The mere fact that the "tensile strength" of water is consistently different in different plant cells indicates that the notion of "designed" leaks may not be so far-fetched. Let us look first at fern sporangia again. Both Ursprung (1915) and Renner (1915) found independently a tensile strength of water in annulus cells between 300 and 350 atm. In a more recent paper, Haider (1954) confirmed this for a number of species of the fern family Polypodiaceae. But in the fern *Aneimia rotundifolia* he found tensions up to 440 atm. *Selaginella* and *Equisetum* have similar spore-ejection mechanisms which snap at lesser tensions (Ziegenspeck 1928). It is obvious that it is not the property of the water, but the structure of the cell wall that must be different. The question then is wheter annulus cells of ferns of the Polypodiaceae have wall perforations of ca. 10 nm diameter and those of *Aneimia* of ca. 7 nm. Renner (1925) and his students Holle (1915) and Frenzel (1929) worked on this problem. Annulus cells are permeable to potassium nitrate, glycerol, but not to sucrose. It was therefore concluded that the pores cannot be much wider than ca. 1 nm, the size of the sucrose molecule. However, permeability even of dead cells depends not only on pore sizes, but also on the respective chemical nature of the wall and the permeating molecule, it is therefore somewhat questionable whether permeability studies can be interpreted so "mechanistically," even though glycerol and sucrose are chemically not too different. Furthermore, pores may well be wider when the cells are under stress (Fig. 3.4 at II).

Let us now explore the question of "designed leaks" further. Renner (1925) was very interested in the nature of cell wall pores and he considered that after a cell's death plasmodesmata deteriorate and the spaces they occupied in the wall would become accessible to air penetration. Holle (1915) and Frenzel (1929) found that air enters dead pith cells of *Sambucus* at a tension of 5–6 atm. This corresponds rather well with the dimensions of plasmodesmata openings. Frenzel (1929) measured pore sizes in all sorts of cells with dye molecules of known size, gold suspensions, etc. and found ranges of pore sizes which become available to air penetration after the death of cells.

3.4 Storage of Water

This section is not meant as a general review of water storage in plants, but concerns primarily storage mechanisms that are more or less directly related to water movement. The following mechanisms come to mind.

1. Elasticity of Tissues. As xylem pressures fluctuate, the plant body shrinks and swells. When pressures drop, water becomes available to transpiration by a volume decrease of the plant body, and when they rise, water is stored. This mechanism is based on osmotic water movement, on imbibition, and on cell-wall elasticity.

2. Capillarity. Any water that is held by capillarity, must be in equilibrium with the apoplast pressure. This can provide considerable storage potential in rigid stems. Strangely enough, it is hardly ever considered.

3. Cavitation. Some plants have special water storage cells, giving up water by "air seeding" at a given tension.

The first mechanism is practically unavoidable and must function at all times, because of the elasticity of the plant body. Osmotic water movement can provide considerable water storage capacity in living tissues. Some plants have special water-storage cells near the photosynthetic cells. Their anticlinal walls crumple when they give up water, but they are still alive and refill when apoplastic pressure rises (Fig. 3.7 A, B). The discovery of diurnal swelling and shrinking is ascribed to Kraus (1881) and his student Kaiser (1879). Leaves, fruits, nuts, and other plant parts shrink and swell diurnally as a result of water-content changes (MacDougal 1924; Tyree and Cameron 1977, etc.). Friedrich (1897) designed a device for the accurate measurement of tree growth and recorded diurnal swelling and shrinking of tree trunks due to water stress. But according to MacDougal et al. (1929, p. 8) his efforts were still much hampered by the high temperature coefficient of the metal out of which his device was constructed. Instruments of this nature, dendrometers or dendrographs, have subsequently been used by numerous investigators, perhaps most extensively by Nakashima (1924) and MacDougal (1924). A more recent description has been published by Fritts and Fritts (1955). A survey of the devices that have been designed over the years has been given by Breitsprecher and Hughes (1975). Part of the expansion and contraction of a stem resides in the bark (Kraus 1877; Klepper et al. 1971), but it has been rather well documented that xylem also undergoes diurnal diameter changes (MacDougal 1924). Even individual vessels contract under stress without rupture of the water columns in them (Bode 1923).

Our discussion here will concentrate on the second and third of the listed mechanisms. The second, capillarity, has been almost entirely ignored in the past, and the third, cavitation, has been much discussed. Let us begin with the rather puzzling question of how tree stems can be relatively "dry" in the summer and "refill" in the fall. By far the most detailed information about this phenomenon has been provided by Gibbs (1958).

In many papers R.D. Gibbs reported the seasonal changes in water content of tree stems. This work was originally started to supply the practical information of

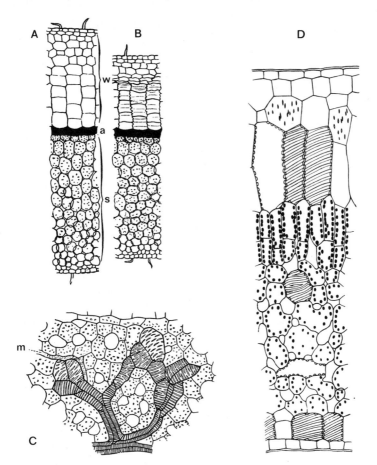

Fig. 3.7. Example of water storage by osmotic water movement, i.e., elastic volume change (**A**, **B**), and by cavitation (**C**, **D**). A and B are water storage cells in a leaf of *Peperomia trichocarpa.* **A** The leaf in fresh condition; **B** severed leaf that has been transpiring for 4 days; *w* water storage tissue; *a* photosynthetic tissue; *s* spongy mesophyll. **C** Bundle ends with storage tracheids in the leaf of *Euphorbia splendens; m* portion of a latex tube. **D** Transverse section through a leaf of *Phyosiphon Landsbergii*, showing storage tracheids. (Haberlandt 1914)

when to float logs to the mill and avoid losses by sinking. But it soon acquired theoretical interest in relation to the mechanisms of water storage and sap ascent. Gibbs (1958) summarized these investigations in the First Cabot Symposium which was held at the Harvard Forest in April 1957. Many aspects of Gibbs' work are interesting, but we need to concern ourselves here only with the controversial question of water storage in xylem. The water content of wood drops throughout the summer and reaches a low value some time before leaf fall. During about a month after leaf fall, some of the tree species appear to "refill" their stems with water, others do not. It is primarily the ring-porous species that do not refill, perhaps because the wide earlywood vessels become embolized in the fall. The timing of cavitation in ring-porous species has never been recorded to my knowledge, although Hermine von Reichenbach (1845) found that tylosis formation in *Robinia* takes place in the fall. As embolism normally precedes tylosis formation (Chap. 6.3), we may assume that at least some vessels embolize in ring-porous trees before winter (Zimmermann 1979).

The interesting and puzzling finding of Gibbs (1958) was that diffuse-porous species showed an increase in water content after leaf fall, amounting to roughly 15% (weight per volume of wood). In *Acer saccharum* and other maple species, it was closer to 25%. This phenomenon has been puzzling because it is hard to conceive that vessels that have embolized during the summer could refill in the fall in the absence of positive pressure. Where is this water storage space?

Some space is provided by the elasticity of the stem, i.e., the mechanism No. 1 above. Some of the tissue contraction is released by cutting while air is drawn into the cut vessels. But this is not enough. In addition it may well be capillarity, a mechanism neglected in the past, although it has been known for very specialized structures. Schimper (cited in Haberlandt 1914) described well-developed intercellular spaces for water storage in petioles of an epiphytic *Philodendron* species. When the external water supply fails, water moves from these spaces to the leaves. The amount of gas space present in wood is not very well known, but it could be considerable. A wood section on a microslide is entirely liquid-saturated, comparable to a log that sinks in the river. But the wood of a living tree is not that much water-logged, it contains a considerable amount of gas. I became first aware of the extensive gas-duct system when we made paint infusions for the vessel-length distribution measurements described in Chap. 1.2. We used paint applicators made of metal (Zimmermann and Jeje 1981, Fig. 1) which were attached to the cut end of the log. The applicator was first filled with destilled water for vacuum infiltration of the cut vessels. Air that had entered the cut vessels after felling the tree and cutting the log, could thus be removed. To our great surprise, water in the applicator kept bubbling no matter how long we applied the vacuum. It was quite evident that air was entering the cut end of the log beside the applicator, was drawn through the intercellular spaces and into the applicator. We do assume that intercellular spaces are necessary to provide ducts for air movement in wood, but it was a surprise that there was so much of it.

Let us now explore capillarity of these spaces with a very much simplified model in order to see how it could work and how effective it might be. The purpose of this exercise is to show how the changing radii of the air–water interface can provide storage space in any crack between solid surfaces. Let us assume that wood consists of bundles of perfect cylinders (representing the fiber matrix of wood). These cylinders can be packed in a tight, hexagonal pattern. Intercellular spaces are probably under atmospheric pressure ($+1$ atm $\cong +0.1$ MPa). Let us assume that the xylem pressure during a summer day is -14 atm (-1.4 MPa) and in the fall -0.5 atm (-50 kPa). In the fall therefore, extracellular water sits in cracks of a width less than 2 μm [see Eq. (3.1)], as illustrated in black in Fig. 3.8 (left). The two large circles represent the cylinders (the fibers) with a radius R of 5 μm. The circle with the radius $r1 = 1$ μm, drawn to scale, represents the water meniscus at -0.5 atm pressure, i.e., the fall condition. In the summer this meniscus is smaller; at -14 atm it is 0.1 μm ($= r2$). On the right of Fig. 3.8 these menisci are shown at higher magnification. During the summer at greater tension, water withdraws to form the smaller meniscus r2; in the fall, as the pressure in the xylem rises, water

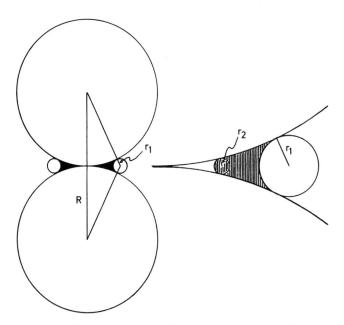

Fig. 3.8. Capillary storage of water (*black*) between cylinders. The radius of the cylinders, representing wood fibers, is shown as $R = 5$ μm, r1 is the radius of the meniscus at -0.5 atm pressure, and r2 the radius of the meniscus at -14 atm pressure. Dimensions are drawn to scale. On the right details are shown at higher magnification. The hatched area is the storage capacity when the pressure increases from -14 to -0.5 atm

(ascending from the roots) must leave the conducting elements to form the larger meniscus r1. The hatched area shows the capillary storage area that is refilled after leaf fall. If the cylinders are tightly packed and endless, there are six such water-storing cracks per cylinder. We can now calculate the dimensions of this space and find that for the above conditions it represents ca. 6% of the total volume. This value may be too optimistic, as capillarity may depend upon hydraulic, rather than measured radii. The purpose of the model is to show the principle of capillary storage. Wood does not consist of perfectly cylindrical cells; potential storage locations have yet to be discussed.

Intercellular spaces are rarely seen between axial elements, although air penetration tells us that they must exist (see, e.g., p. 152 in Baas and Zweypfenning 1979). It appears that there are more spaces along rays than along axial tissue. Bolton et al. (1975) recently reviewed our knowledge about interstitial spaces in wood and described their own observations in the Araucariaceae. MacDougal et al. (1929) made extensive measurements of the permeability to air of intercellular spaces in wood and reported that they are exceptionally tight in *Sequoia*. *Sequoia* may depend less on oxygen in undissolved gas form, it is perhaps this property that makes it more resistant to flooding than other trees. The entire question of aeration in trees was discussed by Hook et al. (1972). The important message of the very much simplified model shown in Fig. 3.8 is that the changing meniscus radii provide very effective water storage in wood, however it does not show the location of the space.

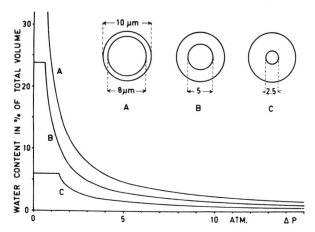

Fig. 3.9. Capillary storage inside dead, air-seeded wood fibers. The fiber is considered to be a 0.4 mm long double cone; storage capacities are shown for three different wall thicknesses (A–C). Δp is the pressure difference between the fiber lumen and the apoplast water. Further explanations in the text

We now have to consider two very important points. First intercellular air spaces are very small in diameter. Vacuum could not pull air through any wet pore with a diameter smaller than 3 µm. If, however, we visualize that the surfaces of these intercellular spaces are partly or entirely lined with a water-repellent coating, it could be done. Second, it is very likely that the largest gas volume in the wood is present inside dead fibers, fiber tracheids and tracheids, and perhaps vessels where the water columns have cavitated (see Fig. 6.1). Vapor pressures in intracellular spaces are anywhere between 0.023 and 1 atm. The slender tips of these cells can store and release considerable amounts of water as the meniscus radii of the capillary water vary. This situation is shown in Fig. 3.9 which illustrates another capillary-storage model, namely a single wood fiber. Its outside diameter is 10 µm, its transverse section circular. For mathematical simplicity it is assumed to be a double cone of 0.4 mm overall length. Three wall thicknesses are considered, corresponding to a central lumen diameter of 8, 5, and 2.5 µm respectively (A, B, and C in Fig. 3.9). The fiber is considered dead and air seeded, i.e., it may contain water vapor. We can now calculate the amount of water it must hold by capillarity at given differences of pressure between the lumen (water vapor) and the apoplast. When the apoplast (xylem) pressure is very low (Δp large), capillary water sits only in the very tip of the pointed fiber lumen. As Δp decreases, the radii of the water menisci increase and water is taken up into the fiber lumen. In Fig. 3.9 the amount of this water is expressed in percent of the total fiber volume (i.e., it would be 100% at Δp = 0 and an infinitely thin wall). The curves show a peculiar flat segment near Δp = 0, this is where capillarity has filled the cell completely. Obviously, once the cell is full, it cannot take up more water. The model ignores complications such as elasticity.

In considering this model, we see that the greatest storage capacity is at small Δp, i.e., at xylem pressures which are not very low. This is exactly what happens in autumn when the leaves drop: xylem pressures increase to values little below atmospheric and the stem is able to store a considerable amount of water. This volume of water, together with the volume taken up by elasticity and imbibition probably provided the water storage described by Gibbs (1958).

We now have to consider a complication. Some fibers may contain water vapor only (and an infinitesimally small amount of air from the air-seeding process), others may contain additional gasses such as carbon dioxide. Gas-filled fibers are known from a number of species, such as maples, apple trees, etc. In this case the inside gas pressure changes according to the gas equation whenever the xylem pressure changes, i.e., Δp may not quite reach zero, even at positive xylem pressures, because the gas pocket in the lumen is merely compressed. In time it may become dissolved, of course.

When pieces of fresh sapwood are vacuum infiltrated and left submerged for a day or so, they take up a considerable amount of water, increasing in weight by 50% or so. This increase is essentially the storage capacity at 1 atm pressure, but of course we know that xylem pressures are rarely that high in a standing tree. However, we do not know how much of this storage is due to elastic expansion (imbibition and osmosis), how much is extracellular storage in cracks between cells, and how much is capillary storage inside vapor-filled cells. Wood contains too many irregularly shaped cells, and does not provide us with neat spherical surfaces for easy mathematical exercises. Milburn reported experimental measurements of water storage capacity in the absence of cavitation with *Fraxinus* leaves. The relative water content of ash leaves decreases from 100% to about 70% under stress with only very few cavitations indicated by clicks (Fig. 3.10). This evidently is capillary and elastic storage. The few cavitations that do take place in this range (say at less than -30 atm) must be irreversible losses of conducting channels. Milburn indeed found that there was no full recovery when leaves were rehydrated (Chap. 3.2).

Capillary storage poses a serious problem with pressure measurements with the pressure bomb (Scholander et al. 1965). Pressure-bomb measurements assume only storage by elasticity (imbibition and osmosis). In respect to these parameters, negative pressure can be substituted by positive pressure. In the case of capillary water storage, this is not possible. Any embolized tracheid, for example, that is *emptied* by negative pressure, is *filled* by positive pressure, unless the applied positive pressure has access to the cell lumen (which it probably rarely has). Even then the pressure difference between the water volume and the air space does not exist, the meniscus radius can therefore not decrease to yield the water! Such errors of pressure-bomb readings have been reported by Boyer (1967) and Kaufman (1968) who compared pressure-bomb readings with psychrometer results. They considered these discrepancies to be caused by the filling of non-xylem tissues and recommended that routine pressure-bomb readings be calibrated for each species by psychrometer measurements.

West and Gaff (1976) ascribed excessive pressure-bomb readings to cavitation of xylem water. While leaves were permitted to transpire after excision and before insertion into the pressure bomb, cavitation clicks could be recorded (Chap. 3.2). Subsequent pressure-bomb readings increased. If the xylem contains large numbers of embolized tracheary elements, pressure-bomb readings can become dramatically increased. Julian Hadley (personal communication) found in March 1982 that in conifers of the timberline in Wyoming, needles of individual twigs gave

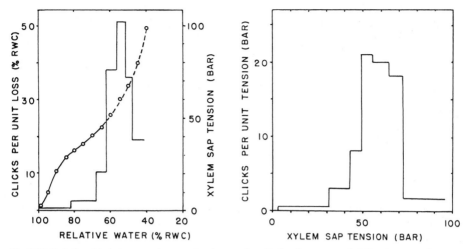

Fig. 3.10. Relationship between xylem-sap tensions and cavitation in manna ash (*Fraxinus ornus*) compound leaf with petiole. *Left* Histogram shows relationship between clicks produced per unit relative water content. Curve shows equivalent xylem tension inferred from pressure bomb extraction of water to give different relative water contents. (*Brocken line* is extrapolated from the pressure-volume curve). *Right* Numbers of the xylem conduits disrupted by cavitation (clicks) as xylem sap tensions increase. Most cavitations occur from 30 to 70 bar tension. (Zimmermann and Milburn 1982, courtesy John A. Milburn)

pressure-bomb readings of − 30 atm on the sheltered side, and exceeded − 70 atm on the exposed side of the same twig, while the entire twig gave a reading of the order of − 30 atm. We are here undoubtedly dealing with wide-spread embolization of tracheids on the exposed side.

When pressure-bomb measurements are made with young twigs, which we can assume contain little embolized water-containing tissue, the error might be tolerable, especially if the measurements are made for comparative purposes.

The third mechanism, water storage in certain cells and retrieval by "air seeding" is probably a very special mechanism that does not play an important role in trees. *Sphagnum* spp. are supposed to store water in special cells which are air seeded at very slight tensions (Huber 1956). Herbs whose xylem experiences positive pressures nightly are another possible case (see below).

Haberlandt (1914) described water-storing tracheids in leaves. They are roundish in shape, and located either at the tip of the veins or even detached from the transporting xylem (Fig. 3.7 C, D). In more recent papers they have been called "tracheoid idioblasts" (Foster 1956; Pridgeon 1982). These could operate in various ways, depending upon their submicroscopic structure, i.e., the dimensions of their wall pores (Chap. 3.3). If air-seeded like ordinary tracheary elements, they could refill completely during a rain, and thus serve as water-storage compartments. In the absence of atmospheric pressure, and thus in the absence of complete refilling, the space between the spiral thickenings and the tips of the cells holds variable amounts of water, depending upon the dimensions and the xylem pressures. The spacing in Pridgeon's illustrations is about 2 µm, a very suitable dimension. In leaf tissue one also encounters occasionally large, branched, almost star-shaped sclereids (e.g., Figs. 5 and 6 in Foster 1956). The shape and dimensions of these are ideal for capillary storage when embolized. In general, air-seeded fibers are ideal for water storage, because their slender tips could hold a considerable amount of water.

I sometimes wonder if tracheids of coniferous leaves do not have "designed leaks" as the above-mentioned observation by Hadley seems to imply. Even under the most severe conditions, tensions exceeding some 30 atm are rarely recorded in coniferous twigs. When stress becomes too great, they dry out and turn brown (for example in trees at the timber line). If "designed leaks" are located at needle tips, water could be "retrieved" into the stem. We might well have here a clue to the function of the bundle sheath in the coniferous leaf. If the bundle sheath in coniferous leaves is open at the distal end, it could favor distal air seeding of the tracheids, and water would be drawn toward the stem, if tensions become excessive. This would be an extreme safety measure, the leaf would thus be sacrificed. But we do know that coniferous needles turn brown and die under extreme stress conditions. Even in dicotyledonous leaves, it is sometimes only the margins that dry up. This is a fascinating area for future research.

When water columns break in regular tracheary elements, they give up water to other plant parts and thus constrict the flow path. Milburn (1973 b) suggested that this might be an important mechanism to regulate water flow and distribution in herbaceous plants that produce positive root pressures and guttate nightly. This would refill the xylem again at night.

Let us briefly summarize the concept of capillary storage of water. The meniscus radius between the cell wall and the gas space is dictated by the pressure difference between the two phases. The difference in space occupied by capillary water at different xylem pressures appears to be quite large. A small part of this may be located in the intercellular space system of wood, mostly along rays, unless these surfaces are water repellent. However, most of the capillary storage space is probably located in the slender tips of embolized fibers and tracheids. It is important to realize that all this space is a single compartment for water, but not for gas. Gas can only circulate in the intercellular space system; in embolized fibers and tracheids, gas is isolated. These separate intracellular spaces (fiber and tracheid lumina) are probably mostly water vapor-filled, because boiling a piece of wood, vacuum infiltrating, or merely soaking it, fills most of this space. Unless there are 3 μm-wide holes in the cell walls of embolized fibers, it is impossible to remove gas from them by vacuum infiltration [Eq. (3.1)].

The fact that xylem water pressures are usually below atmospheric give the land plant two advantages. First, in case of injury, the system is self-sealing (see the next section). Second, negative pressures provide embolized xylary elements for capillary storage as we have seen. But what happens in plants where pressures are perennially positive as in aquatic angiosperms? Of course the plant needs a separate air duct system and has to seal it off the apoplast! We shall have to come back to this in Chap. 5.4.

3.5 Sealing Concepts

The direction of flow in transport systems is normally regulated by demand. If the pressure drops at the receiving end, flow intensifies. It is possible that an accidental sink, caused by an injury, drains the system and thus represents a grave danger. Animals can bleed to death, but there is a blood-clotting mechanism preventing such a fatal result most of the time. Injured phloem or xylem could be drained by exudation, but there are mechanisms that can seal the injured system. Those of the phloem have been described by Eschrich (1975).

Xylem pressures are rarely above and far more commonly below atmospheric. In case of an injury suffered while pressures are less than atmospheric, the injury is not a sink but a "source" (to use the terms of phloem physiology). Air enters the xylem, displacing the liquid, until the air–water interface has reached a wet membrane, namely the tracheary wall, where its movement is stopped automatically by the surface tension of the water. Even though water flow is stopped, the injured vessel is now open to the outside world and represents a potential entry site for harmful microorganisms. Plants have therefore evolved secondary means of sealing the air-blocked vessel with gums (carbohydrates) or living cells (tyloses) (Chap. 6.3). In addition, injured xylem is often isolated by suberization of the cell walls next to the injury. This seals the extra-fascicular pathway between the damaged area and the living plant part. These secondary sealing mechanisms will be discussed in Chap. 6.3. Let us now return to the first seal, which is provided by capillarity.

Occasionally, pressures in the xylem are above atmospheric. This happens in some species in early spring as mentioned above. It also happens in certain herbs during nights when absorption conditions are favorable and transpiration is minimal. Excess water is then bled from hydathodes at the margins of leaf blades. If the plant is injured when pressures are above ambient, it bleeds. This does not appear to be harmful, because it only happens when there is a surplus of xylem sap. The rather special case of aquatic angiosperms will be discussed later (Chap. 5.4). Most of the time, pressures are below ambient; when the xylem is injured, air is drawn in, it is thus self-sealing by its very structure. We could even speculate that low operational pressures were selected for during evolution, because they not only make xylem self-sealing, but also provide an intercellular air-duct system, by decreasing capillary meniscus radii.

In certain plant parts we find that the cell walls are suberized and free apoplastic water movement thus blocked. This barrier is usually called a Casparian strip. The endodermis of the root with its Casparian strip has been known since the last century. Older literature has been reviewed by Haberlandt (1914). Clarkson and Robards (1975) summarized more recent papers and describe the histologic development of the Casparian strip. The aspect that concerns us here is the fact that the cell walls of the endodermal layer in the root provide an effective seal for apoplastic water movement in radial direction. "Outside" and "inside" apoplast are separated into different compartments by the barrier. This means that pressures in the two can be substantially different. This pressure seal concept was experimentally tested almost 100 years ago (de Vries 1886). Positive pressures in the xylem of the root system would not be possible without such a seal. Transport between the two compartments is metabolically controlled and goes via symplast, i.e., the plasmodesmata (Läuchli and Bieleski 1983). The endodermis is mechanically well developed in plants of dry areas and in marsh plants that dry up periodically (Haberlandt 1914). In extremely dry desert habitats, for example, the Casparian strips are much wider, presumably because the pressure difference between the soil and the stele is greater (Fahn 1964) (see also Chap. 4.1).

Xylem pressures are normally highest in the roots and rhizomes, here they are more often above atmospheric than anywhere else. As pressures in aerial parts are normally below-atmospheric, an apoplastic seal is only necessary at the plant surface in order to prevent water loss by evaporation. This is provided by the epidermis or the periderm. But in desert shrubs, suberization can be found in the stem phloem parenchyma (Jones and Lord 1982), the rays, between xylem layers, etc. (Fahn 1974), thus sealing the xylem more effectively against water loss.

Vascular bundles of some leaves are surrounded by a bundle sheath, containing a suberized layer comparable to that of the Casparian strip in the roots (Schwendener 1890; O'Brien and Carr 1970). This seal separates the apoplast into two compartments, one inside and the other outside the bundle sheath. The two areas are only connected by plasmodesmata that connect living cells (O'Brien and Carr 1970). It may well be that the bundle sheath and/or presence of hydathodes prevent flooding of the photosynthetic tissue in those plants that experience positive root pressures regularly. That flooding can actually happen was found by Stahl (1897), who grew plants that lack hydathodes in warm, humid soil and covered them with glass jars to prevent transpiration. The intercellular space of the photosynthetic tissue was thereby infiltrated. These intercellular spaces are often protected by the poor wettability of their interior cuticle (Häusermann 1944; Scott 1950).

Sperry (1983) studied water relations of the Mexican fern *Blechnum lehmannii*. Measuring root pressure on petioles with bubble manometers, Sperry recorded xylem pressures up to about 0.2 atm above ambient in the plants growing near the water, and progressively lower pressures in plants growing higher up on the stream bank. Only the hydathodes of the young leaves guttated (in intact plants) those of older leaves were sealed and did not guttate. This appears to be a mechanism to direct the (nutrient-containing) xylem stream toward the younger leaves when transpiration is low and positive root pressure exists. We shall encounter this mechanism once more when we discuss aquatic angiosperms (Chap. 5.4).

In a previous section (Chap. 3.3) we looked at the quantitative properties of the seal provided by a wet membrane. Smaller vessel-wall pores of desert plants seal outside air against greater negative pressures (e.g., Scholander et al. 1965). It has been discussed in Chap. 3.4 that air leaks in tracheary elements may be provided by the wall perforations of plasmodesmata left behind after protoplasmic degradation. It is in this respect perhaps of interest to consider Braun's (1963, 1970) *Hydrosystem*. Braun recognizes five levels of evolutionary development of xylem. The lowest level is that of gymnosperms where most of the axial elements are dead tracheids. In the next higher level, some vessels are embedded in a matrix of dead tracheids, fiber tracheids, and fibers. Further evolutionary development gradually surrounded the vessels with (living) parenchyma. In the wood of the most advanced stage, vessels are entirely enveloped by paratracheal parenchyma (Fig. 3.11). This is certainly a very effective seal which protects the vessels from accidental air leaks.

Sealing mechanisms of coniferous tracheids deserve special attention. Tracheids are connected by bordered pit pairs which are located primarily (but not exclusively) on radial walls. Their peculiar structure is shown by one of Bailey's

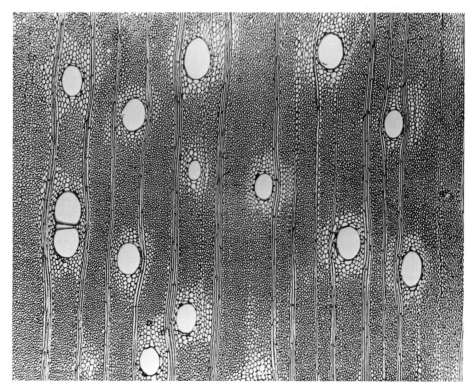

Fig. 3.11. Transverse section of the wood of *Prosopis juliflora* (Leguminosae), showing vessels surrounded by living parenchyma cells (paratracheal parenchyma). The parenchyma cells can be recognized by their lighter color, the darker-colored (i.e., thick-walled) cells are fibers. (Photomicrograph by I.W. Bailey)

(1913) drawings which is reproduced in Fig. 3.12. Two features characterize the coniferous bordered pit, (1) the pit membrane is rather porous, and (2) it contains a central thickening, the torus. Many electron micrographs have been published during the past 30 years, they have been obtained by many preparation techniques such as thin sectioning parallel to the wall to remove the pit border, carbon replication, etc. They all show the pit membrane as a rather drastic modification of the original primary wall. It looks as if the torus is suspended by more or less radially oriented bundles of microfibrils (Fig. 1.11) (see also Frey-Wyssling and Bosshard 1953; Frey-Wyssling et al. 1959). The gaps between this fibrillar network appear to be rather large, of the order of a fraction of a micrometer. Functionally, this means that the membrane alone could not contain an embolus against great tensions. The suspicion could arise that this structure is an artifact of preparation, but permeation experiments by Liese and Bauch (1964) showed that the gaps are real. When wood of various species was perfused with metal suspensions, particles up to 0.2 µm diameter often filtered through. This implies that an embolus could spread in coniferous wood at tensions of only 15 atm. However, the torus is easily displaced if the pressure difference between the two tracheids becomes significant. When a tracheid is injured, its lumen is filled with air at ambient (+1 atm) pres-

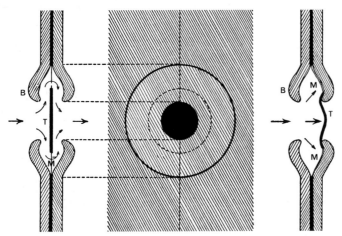

Fig. 3.12. *Center* Surface view of the radial wall of a coniferous tracheid, showing a bordered pit. *Left* The same pit in section, *arrows* indicating the path of water from one tracheid into the next. *Right* Section showing the valve-like action of the torus. *T* torus; *M* pit membrane; *B* pit border. (Bailey 1913)

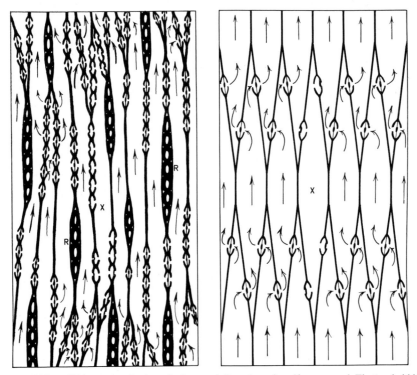

Fig. 3.13. *Left* Diagrammatic view of a (tangential) section of coniferous wood. The tracheid in the center (marked *X*) is vapor-blocked and does not function; water flows around it. The negative pressure in the conducting tracheids has pulled the pit membranes away from the vapor-blocked cell to seal it off. Note that in reality tracheids are much longer in relation to their diameter. *Right* The same situation is shown in a still more diagrammatic manner. Note that coniferous tracheids are never exactly storied as this diagramm may imply. (Certain dicotyledonous woods have their cells storied, i.e., arranged all at the same height like books on a book shelf)

sure. The pressure in the intact, water-containing neighboring tracheids may still be negative; a considerable pressure drop therefore exists across the pit membranes. Their tori are therefore pulled against the pit borders of the neighboring, water-containing tracheids (Fig. 3.12, right). The air-blocked tracheid is thus sealed off from the water-conducting tracheids not only by capillarity, but also by the valve action of the torus. Dixon (1914) showed this situation rather nicely in his Figs. 17 and 18. Tracheids are so much longer than wide that a naturalistic drawing would not be very clear. Figure 3.13 is an attempt to explain the situation at two levels of simplification.

How well does the valve action of coniferous bordered pits work? The question is particularly interesting in the cases where the surface of the tertiary wall of the tracheids contain warts. Krahmer and Côté (1963) published such electron micrographs (e.g., their Fig. 11), and the seal looks quite effective. Nevertheless, pressures in coniferous xylem rarely drop below -30 atm, in other words, we must assume that the seal fails at this level. This indicates residual leaks of around 0.1 μm diameter.

An interesting indication of the function of coniferous bordered pits was reported by Hudson and Shelton (1969). When they pushed liquid through freshly cut stems of southern pines, they could increase the flow rate by a factor of 20–30 by first cutting off a 3–5 cm thick disk from the application end. When, after initial perfusion, they cut off a second disk, the flow rate increased 400-fold. This can be explained as follows: After felling, the xylem of the stems was still under tension, the bordered pits at the cut ends therefore closed. This made perfusion difficult. Successive removal of these sealed xylem regions improved xylem conductance, particularly when the stem xylem was first pressurized, i.e., the tension released.

Finally, most conifers possess a mechanism to seal off injured xylem parts by resin impregnation. Even those species which do not normally have resin ducts in the wood, develop traumatic ducts upon injury. Similar mechanisms exist in some dicotyledons. This seal is very effective in that it closes the entire apoplast including the cell walls, a phenomenon which will have to be taken up in Chap. 6.3.

3.6 Pressure Gradients

Positive xylem pressures are relatively easy to measure. A simple and convenient device for this is the bubble manometer. It consists of a glass or transparent plastic capillary, partly filled with liquid. The far end of the capillary is closed and contains air. The near end may terminate with a hypodermic needle, it is liquid filled and is brought into tight contact with the liquid whose pressure is to be measured. It has been quite useful to measure sieve-tube pressure (Hammel 1968; Sovonick et al. 1981). To measure positive xylem pressures one may drill a small hole into the wood and stick the tightly fitting needle into it. The xylem pressure compresses the air pocket at the far end of the capillary. This air volume can be read off the scale as a length. As pressure times volume is constant (at constant temperature) xylem pressure can easily be calculated. Corrections for temperature can be included if necessary if the manometer is to be left in place for extended periods. One has to be aware of the fact that the initial reading before insertion is the ambient

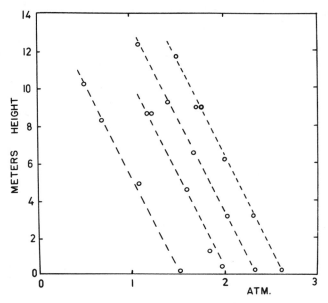

Fig. 3.14. Gradients of positive pressures (in atm above ambient) in the xylem of stems of grapevine in the spring before the buds open. (Scholander et al. 1955)

barometric pressure, in some cases this may also have to be taken into account when calculating pressures. Furthermore, readings can become unreliable when the pressure drops much below atmospheric. Air pressures of the intercellular space and water vapor pressure may become significant. Many early pressure measurements with such devices are therefore useless. Nevertheless, the bubble manometer is ideal for the measurement of positive pressures. Scholander et al. (1955) used such a device to measure gradients of positive pressures in grapevine during early spring before bud opening. They found gradients that conformed perfectly to the 0.1 atm m^{-1} hydrostatic gradient (Fig. 3.14).

Manometer measurements become unreliable when pressures are much below atmospheric, and they are totally useless when pressures are negative. Unfortunately this is the case most of the time. But the introduction of the pressure chamber by Scholander et al. (1965) made measurements of gradients of negative pressures along a tree trunk possible. Like everything, the pressure chamber has its limitations. Measurements cannot be taken on the stem xylem directly, one needs a small twig or leaf. One also must realize that as one applies outside pressure to a twig, one may fill the capillary space system discussed in Chap. 3.4, because the pressure difference between the intercellular air and the xylem water decreases, thus increasing the interface meniscus radii. Boyer (1967), Kaufman (1968), West and Gaff (1976) have given this problem special attention.

Figure 3.15 shows pressure gradients that were obtained by Scholander et al. (1965) on Douglas-fir. The twigs were obtained by shooting them down with a rifle. The gradient shows the expected drift into a more negative region during the daylight hours at the time one would expect transpiration to intensify. What is surprising, however, is the fact that the gradient is always close to 0.1 atm m^{-1}, indeed

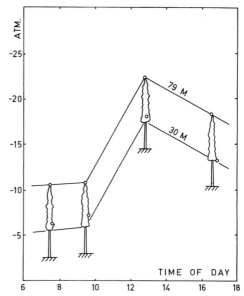

Fig. 3.15. Gradients of xylem pressures in Douglas-fir. (Redrawn from Scholander et al. 1965)

often less. This is physically impossible if all our assumptions are correct. During the mid-day hours the gradient should be steeper on account of flow resistance. This problem was investigated by Tobiessen et al. (1971) with a great deal of care. They mounted an elevator on a tall redwood tree and took shaded twig samples close to the trunk, on a cool (11°–14° C) and relatively humid (70%) day. They sampled at five levels from ca. 20 to 80 m height and found a gradient of only 0.08 atm m^{-1}, i.e., less than hydrostatic. When they corrected the pressure values to account for extracellular water (Chap. 3.4), the gradient even decreased slightly. Gradients measured with the pressure chamber do sometimes appear steep enough to account for both gravity and resistance to flow. A good example are those Connor et al. (1977) measured in *Eucalyptus regnans*. It is possible, of course, that even these show a diminished slope.

The solution of this problem is relatively simple. There are pressure gradients up the trunk and out into the branches. When one samples twigs, one goes quite literally out onto a limb. In other words the recorded pressure may be much lower than the pressure at the same height in the stem. Hellquist et al. (1974) and Richter (1974) drew attention to this problem. As we shall see in Chaps. 4.3 and 4.4, the xylem path may be quite constricted at branch junctions, and the conductivity of branches is less than that of the stem. Pressure chamber measurements cannot be taken as pressure values of the stem xylem without special precautions, simply because they are taken elsewhere!

There is a relatively easy way to remedy the problem. The twigs to be measured can be identified and their transpiration can be suppressed by enclosing them in plastic bags. At the same time one has to cut off all leafy parts of that branch so that transpiration in that lateral branch is entirely suppressed. It would be interesting to repeat the measurements of Fig. 3.15 in this way, but this has not yet been done so far as I know. All we can say at this point is that pressure gradients in tall

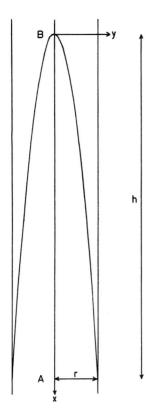

Fig. 3.16. The flow paraboloid in a capillary of the radius *r*. If, at zero time, we could label all water molecules on a transverse-sectional plane at **A**, they would be lined up on the surface of a paraboloid at time *t*. The fastest ones, in the center of the capillary, would have covered the distance *h* and reached the point *B*. A parabola is usually shown as y^2 = 2px; the x and y axes are indicated, but the drawing would to be rotated 90° counterclockwise

trees seem to be of the order of 0.1 atm m^{-1} (in addition to the gravity gradient). They may be steeper in small trees and can be very considerable in herbaceous plants as will be briefly discussed in the next section.

A more recent device for measuring stem pressures is the aquapot, an osmotic tensiometer (Peck and Rabbidge 1969). Legge (1980) obtained a large number of very interesting pressure gradient recordings on tall eucalyptus trees. They even include reverse gradients indicating downward flow following a rain. The aquapot can be inserted into the stem and left there. The only reservation one might have is that it has not been critically calibrated with other instruments with whose performance we are familiar.

3.7 Velocities

Flow through ideal capillaries is paraboloid as described in Chap. 1.3. Let us now visualize this in more detail with the aid of Fig. 3.16. The capillary shown here has the radius r. Imagine that we could label all water molecules of a transverse-sectional plane at A. We then let them flow for the time t. The molecules have moved at different velocities, they are now all spread over the surface of a paraboloid. The ones in the center moved fastest, they have reached the point B and covered the distance h. The volume of the paraboloid is r^{2}πh/2, whereby r is the capillary radius

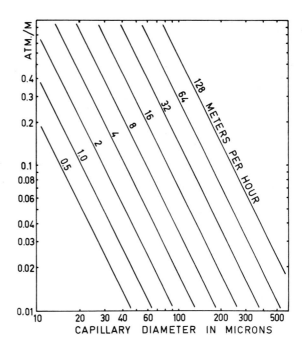

Fig. 3.17. The relation between pressure gradient, capillary diameter, and flow velocity. (Zimmermann and Brown 1971)

and h the height of the paraboloid. This volume is equal to the volume flowing during a given time t whereby the height h is proportional to the time during which flow has taken place. The flow rate, according to Hagen-Poiseuille, has been given in Eqs. (1.2) and (1.3). We can set this equal to the volume of the paraboloid

$$\frac{Q}{t} = \frac{r^4\pi}{8\eta} \times \frac{dP}{dl} = \frac{r^2\pi h}{2t}, \tag{3.2}$$

whereby Q/t = the volume flow rate, η = the viscosity, and dP/dl the pressure gradient. The peak velocity is therefore

$$\frac{h}{t} = \frac{r^2}{4\eta} \times \frac{dP}{dl}. \tag{3.3}$$

Velocity is directly proportional to the pressure gradient and to the square of the capillary radius (Fig. 3.17). The fact that flow follows a paraboloid is important conceptually because it means that velocities range from zero at the vessel walls to a peak in the center of the vessel. When measuring velocities and when talking about them, we must be aware of this. Peak velocity is a useful concept which we shall discuss further. Another concept is average velocity. This is based upon the notion that liquid moves like a solid cylinder. This never happens, of course, but it is nevertheless a useful concept when dealing, for example, with specific mass transfer in the sieve tubes (Zimmermann and Brown 1971). It like to refer to this as "cylindrical velocity," a term that is not easily misunderstood.

When we measure velocity, we usually try to detect the first arrival of a solute (a dye, an isotope, etc.), or perhaps a heat pulse that has been introduced into the flowing water upstream, as will be discussed later. The first arrival of this "signal" gives us information about peak velocity, i.e., the velocity in the center of the largest vessels. However, our detection sensitivity is limited. It is quite obvious from Fig. 3.16 that the peak gets more slender, the farther is moves. If we had, for example, a certain tracer concentration at A that we call 100%, the concentration at distance h (at the very peak) is zero. Ten percent behind the tip, the concentration is 10%, 20% behind the tip 20%, etc. The shape of the paraboloid just happens to give us this straight-forward relationship. In other words, whatever we measure, we never quite get the peak, hence we are always dealing with minimum peak velocities. Another reason why our measured velocity may be somewhat too small is the fact that over long distances the tracer may get lost by lateral diffusion.

Velocities can be measured with tracer (dye or isotope) only on cut shoots after the cut end has been immersed in water and xylem pressures are relaxed. If the tracer is injected into an intact plant, a point of $+1$ atm is introduced into the xylem stream and the existing pressure gradient is thereby totally changed. This can be seen from the fact that injected dyes always move down as well as up in the stem. Acid substances move fastest, alkaline substances get absorbed on the wall. The latter are useful to mark a track, but cannot be used for getting even a rough velocity measure.

The most elegant method to measure flow velocities is the thermo-electric or heat-pulse method. It was first used by Rein (1928) to measure flow velocities of blood in animals, and adapted for the measurement of xylem sap velocities by Huber (1932). The principle is to heat up the liquid briefly at a point along the stem and time the arrival of the heat pulse downstream. A little wire loop underneath a bark flap is heated up by an electric current of a very brief duration. The arrival of the heat wave is recorded by a thermocouple 4 cm downstream from the heating wire. With this arrangement Huber and Schmidt (1936) made many measurements on cut branches and on standing trees and reported diurnal as well as annual flow velocity changes. But this arrangement does not work well with velocities less than 1 m h^{-1} (0.3 mm s^{-1}). For such slow movements (for example in the xylem of conifers), Huber and Schmidt (1937) developed the so-called compensation method in which one of the thermocouple junctions is located 16 mm upstream, the other 20 mm downstream from the heat source.

Methods used today sometimes differ slightly in configuration (in some procedures a small hole in drilled into the wood, thus interrupting the xylem at that point) and one has to be careful when trying to obtain absolute values (Marshall 1958; Swanson and Whitfield 1981; Cohen et al. 1981). In fact, absolute velocity measurements are quite useless (see Chap. 4.3). Velocities range from zero at vessel walls to a peak in the center of the widest vessel. Vessel diameters usually vary quite a bit in any plant part. The only well-defined velocity is therefore the peak velocity in the center of the widest vessels, but it is doubtful if this is ever recorded (Rouschal 1940, p. 230). In addition, ecologists are often not really interested in velocities but in volume flow that can be calculated from heat-pulse measurements with proper calibration.

The heat pulse method has given us a great deal of *comparative* information about xylem sap velocities in plants. Velocities decrease to almost zero at night and usually reach a peak shortly after noon. The curves shown in Fig. 3.1 show three major points. First, velocities in wide-vessel trees like ash are considerably greater than those of narrow-vessel trees like birch. Second, in some trees (e.g., ash) velocities diminish from bottom to top of the trunk, and in others (e.g., birch) they increase. Third, the diurnal plots of velocities at the bottom vs. top of a tree show a hysteresis loop which indicates that water is pulled up by transpiration (Chap. 3.1). Huber and Schmidt (1936, 1937) also reported peak velocities of many different species. Trees with wide vessels (diameters of 200–400 µm) showed mid-day peak velocities of 16–45 m h^{-1} (4–13 mm s^{-1}). Trees with narrow vessels (50–150 µm) had slower mid-day peak velocities, ranging from 1 to 6 m h^{-1} (0.3–1.7 mm s^{-1}). To this latter group belong two species of the dry Mediterranean region, namely *Fraxinus ornus* and *Quercus ilex*, whose northern relatives belong to the first (wide-vessel) group (Rouschal 1937). Slowest peak velocities were recorded on conifers with the compensation method; they corresponded to the slowest velocities of narrow-vessel trees.

It is useful to see what pressure gradients are required to achieve such mid-day peak velocities (Fig. 3.17). Even if we consider that vessels are only about 50% efficient (Chap. 1.3), it can be seen that pressure gradients in tree trunks, due to flow resistance, should be of the order of 0.1 atm m^{-1}. This seems to be generally true for trees. But consider now a very short herbaceous plant. If conditions in leaves and roots are similar to those of trees, then the plant can afford very much steeper gradients and very much greater velocities. Begg and Turner (1970) found pressure gradients of 0.8 bar m^{-1} in tobacco. Rouschal (1940) found a maximum velocity of 108 m h^{-1} in *Impatiens parviflora*. More recently Passioura (1972) found in wheat plants that flow rate measurements indicated velocities up to 900 m h^{-1} (25 cm s^{-1}) through a single vessel of the tap root. By forcing plants to obtain water from a single root, velocities could be increased experimentally to 2,880 m h^{-1} (0.8 m s^{-1}). Comparable findings (i.e., steep pressure gradients in small plants) were reported by Woodhouse and Nobel (1982).

Chapter 4
The Hydraulic Architecture of Plants

4.1 The "Huber Value"

As a painter and naturalist, Leonardo da Vinci was a keen observer of nature. He made the following entry in his notebook about the construction of trees:

"All the branches of a tree at every stage of its height when put together are equal in thickness to the trunk (below them). All the branches of a water (course) at every stage of its course, if they are of equal rapidity, are equal to the body of the main stream. Every year when the bows of a plant (or tree) have made an end of maturing their growth, they will have made, when put together, a thickness equal to that of the main stem; and at every stage of its ramification you will find the thickness of the said main stem; as i k, g h, e f, c d, a b, will always be equal to each other; unless the tree is pollard – if so the rule does not hold good." (Fig. 4.1) (Leonardo's notes 394, 395, Richter 1970).

Botanists are well aware of the approximate correctness of this statement; measurements were made around 1900 to investigate the significance of stem dimensions in satisfying both mechanical and hydraulic demands (Metzger 1894, 1985; Jaccard 1913, 1919; Rübel 1919). This concept of even conductance throughout a tree has been called the "pipe model" by Japanese workers (Shinozaki et al. 1964), because a tree can be imagined of consisting of many thin, tall plants, bundled together.

It was a considerable conceptual step forward when Huber (1928) measured transverse-sectional xylem areas of stems and branches and expressed them per fresh weight of leaves that were supplied by that part of the axis. For convenience, I like to refer to this as the "Huber value." The advantage of relating transverse-sectional area to supplied leaf mass is that measurements taken at different points in a plant, or in different plants, become directly comparable. Huber avoided the complication of heartwood formation by working primarily with young trees or tops of older ones. In some cases, he measured the most recently formed growth ring separately. He was also aware of the fact that transverse-sectional area alone was not sufficient to describe hydraulic properties, he therefore measured, in some cases, hydraulic conductivity separately. This automatically removed the disturbing effect of non-conducting xylem such as heartwood.

In addition to the innovation of relating the xylem to the supplied leaf mass, Huber (1928) made two major contributions. His measurements permitted a comparison of axes within an individual tree. An example of this is shown in Fig. 4.2 which illustrates the top of a young *Abies concolor*. The average transverse-sectional area in most part of the tree were around 0.5 mm^2 g^{-1} fresh weight of supplied needles. But this number increased sharply upward along the main stem to a maximum of 4.26 at the base of the leader. If only the most recent growth was measured, the increase was even more pronounced (numbers in parentheses in Fig. 4.2). The

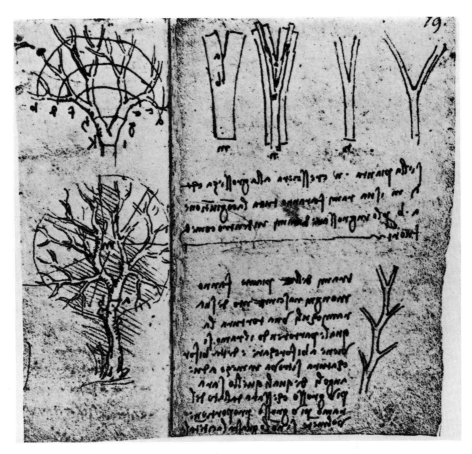

Fig. 4.1. Leonardo da Vinci's sketch of tree architecture showing that the transverse-sectional area of the trunk is equal to the sum of branch transverse-sectional areas. Note that Leonardo wrote in mirror image. (Richter 1970)

implication of this construction is obvious: the leader is better supplied by xylem transport than the laterals. This was a previously unsuspected expression of apical dominance. The second contribution was a comparison of species. The Huber value of stems and branches of dicotyledons and conifers of the north temperate region is about 0.5; while that of plants of more humid habitats is considerably lower. Herbs of the forest floor show an average value of 0.2 (extremes 0.01 and 0.8), and aquatic angiosperms (Nymphaeaceae) 0.02 (Gessner 1951). Plants of dry habitats have much higher Huber values, the average for Egyptian desert plants was found to be 5.95 (extremes 1.4 and 17.5) (Stocker 1928; Firbas 1931 b). These latter values exclude succulents which depend heavily upon water storage, whose xylem is therefore not directly comparable with that of other plants. Interestingly, plants of raised bogs have Huber values rather comparable to desert plants (Firbas 1931 a), probably because such plants experience periodic droughts (see also Chap. 4.1).

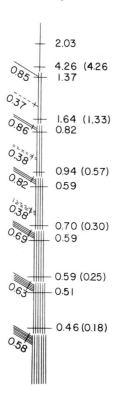

Fig. 4.2. Diagram of the top of a young *Abies concolor*. The age of each segment is shown by the *number of parallel lines*. *Printed numbers* indicate transverse-sectional areas of xylem in mm²g⁻¹ fresh weight of supplied leaves. *Numbers in parentheses* concern the most recent growth ring alone. (Huber 1928)

Filzner (1948) computed xylem transverse-sectional area per supplied surface area of *Rhynia*, and arrived at a figure of about 1.4 mm² dm⁻². *Rhynia* had, of course, no leaves. In dicotyledonous trees one gram fresh weight corresponds to 1–2 dm². The value obtained for *Rhynia* by Filzner is therefore somewhat high, i.e., comparable to values found in desert plants. Differences found in different plant groups may well be based primarily upon tracheary diameters. Plants in dry habitats have generally narrow vessels; it takes many more narrow elements (i.e., a larger transverse-sectional area) to move a comparable amount of water. In the case of *Rhynia*, the pits connecting the tracheids may still have been very inefficient. Filzner's measurements were made on a single 2.5-mm-long piece; it would certainly be worth while to make more detailed investigations on early land plants.

There are reasons other than hydraulic considerations why investigators studied the relationship of leaf mass and wood production. Burger (1953), for example, was concerned with the amount of foliage that is required in different forest types to produce a certain volume of wood. More recently, ecologists have become interested in finding ways to estimate foliage quantity by non-destructive means. Thus Grier and Waring (1974) established the ratio of sapwood transverse-sectional area to leaf dry weight of three coniferous species. This information enabled them to estimate the total weight of leaves of a tree from the dimensions of an increment core. Their results range from 1.3 to 2.1 mm² g⁻¹ *dry* weight of supplied leaves, values quite comparable to those reported by Huber (1928).

4.2 Leaf-Specific Conductivity

Huber (1928) was well aware of the fact that transverse-sectional xylem area is not a good hydraulic measure. He therefore measured also hydraulic conductivity of woody axes as earlier investigators had done (e.g., Ewart 1905–1908; Farmer 1918). By multiplying the xylem area by the conductivity, he was able to obtain a conductance based upon the fresh weight of supplied leaves. He called this Gesamt-leitvermögen, which translates into something like "effective conductivity." This is an informative measure; the only problem with it is that it is an extremely cumbersome way of obtaining meaningful results; it has, in fact, rarely been used. Much more direct measurements are those of hydraulic conductance of an axis expressed per fresh weight of supplied leaves. We refer to this as "leaf-specific conductivity" (Zimmermann 1978a).

Leaf-specific conductivity (LSC) as it is defined at present is based upon fresh weight of leaves. It would perhaps be of advantage to base the conductance measurements upon leaf surface area rather than fresh weight. But if we are dealing with reasonably large trees, i.e., very large quantities of leaves, it is much more difficult to measure leaf areas than fresh weight. Another objection to fresh weight is the possibility of its dependence upon xylem pressure (Fig. 3.10). However, if measurements are done all at one time, for example early in the morning (in order to finish the entire analysis of a tree during the day), and particularly if the prime concern of the measurements is to study LSC distribution within an individual tree, accuracy of absolute values is of minor importance. Leaf-specific conductivity is the amount of water flowing through a piece of stem in microliters per hour, under conditions of gravity flow, per gram fresh weight of leaves supplied by that stem section. These dimensions have been chosen entirely for convenience. They are easily visualized and measured, and they give a range of convenient numbers from about 1 to 1,000. Anyone who has the desire to make them look more scientific may want to convert them to SI units.

In making conductance measurements on pieces of stem, one often runs into the problem of more or less sharp conductance drops over time. This was first reported by Huber and Merz (1958) for coniferous wood, and ascribed to aspiration of bordered pits because they used rather large pressure gradients and found that the conductance drop could temporarily be reversed by reversing the flow direction (one of their figures is reprinted in Zimmermann and Brown, 1971 as Fig. IV-20). When the conductance drop was also found in dicotyledonous wood whose bordered pits lack tori, it was ascribed to tiny air bubbles (and other suspended particles) that settle on the intervessel pit membranes and thereby obstruct the flow (Kelso et al. 1963). Working with dicotyledonous wood, we found that the conductance drop could be almost entirely avoided by using a dilute salt solution rather than distilled water (Zimmermann 1978a). The reason for this is not yet known, but we suspect that intervessel pit membranes are swollen and therefore less permeable when saturated with distilled water. With most species (and specimens) 5 or 10 mM KCl gives maximum and stable flow rates. Larger stems often require higher concentrations (in some cases we used 100 mM KCl). The KCl effect is absent in coniferous wood (*Thuja occidentalis*, Chen et al. 1970, *Pinus radiata*, R.E. Booker, personal communication, *Tsuga canadensis*, tested in our own laboratory).

This is not surprising because the margo of the coniferous bordered pit appears to contain little swellable matrix material and looks on electron micrographs more like an inert screen (Fig. 1.11). Short-term flow-rate measurements are reasonably stable, but long-range flow-rate measurements show the conductance drop one would expect from the accumulation of suspended particles on the pit membranes. The greatest difficulties so far have been encountered with the palm *Rhapis excelsa*, where the conductance drop is very sharp even if all precautionary measures are taken. Membranes of intervessel pits (Chap. 2.2) appear to be particularly easily plugged here.

4.3 The Hydraulic Construction of Trees

Leaf-specific conductivities have, so far, only been published for diffuse-porous tree species (Zimmermann 1978 a). This was a matter of convenience. Ring-porous species are more difficult to deal with, because one may unintentionally measure the additional conductance of old, non-functional vessels. This danger is absent in those species that produce tyloses profusely. Conifers are also difficult to deal with. First one has to select a species without resin ducts in the xylem which could disturb the result, and second, picking coniferous leaves off a tree can be very time-consuming. But lower hydraulic conductivity (conductance per transverse-sectional area of xylem) of branch when compared with trunk wood has been recognized in *Tsuga canadensis* (Tyree et al. 1975).

Figures 4.3 to 4.5 show some results. Let us first compare LSC of main stems. In *Populus* (Fig. 4.3) and one of the *Acer* specimens (Fig. 4.5, left) the values fluc-

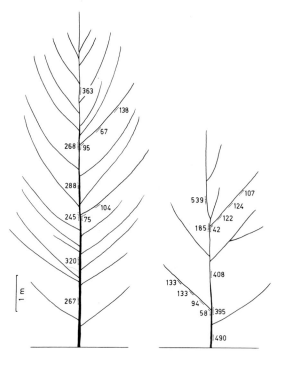

Fig. 4.3. Leaf-specific conductivities along the axes of two open-grown large-toothed poplar trees (*Populus grandidentata*). (Zimmermann 1978 a)

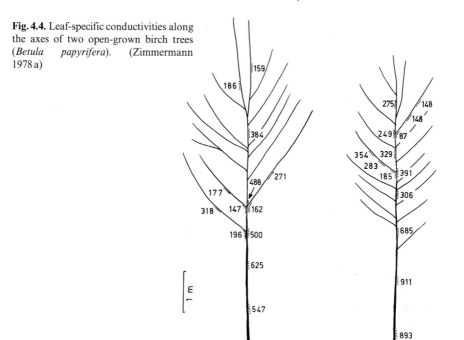

Fig. 4.4. Leaf-specific conductivities along the axes of two open-grown birch trees (*Betula papyrifera*). (Zimmermann 1978 a)

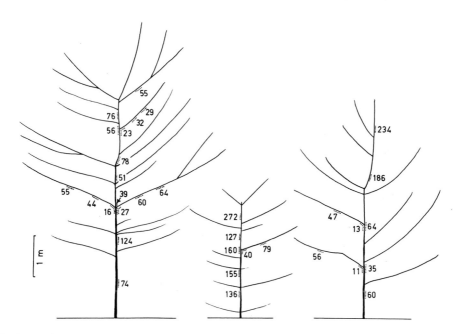

Fig. 4.5. Leaf-specific conductivities along the axes of three sugar maples (*Acer saccharum*). (Zimmermann 1978 a)

tuate somewhat along the trunk, but do not seem to show any particular trend. In the two younger *Acer* specimens, however, there is a distinct increase in LSC from the stem base to the top (Fig. 4.5, center and right), and in *Betula* (Fig. 4.4) there is a distinct decrease in distal direction. Leaf-specific conductivities of lateral branches are consistently smaller than those of the main stem, i.e., about half as large. Finally, junctions show often the lowest LSC values along any specific path of the water.

Measurements of LSC were also taken in 1-year-old and current-year twigs and in petioles. Not as many measurements were made as with stems and branches, because at very slow flow rates, conductance measurements require somewhat different techniques. On petioles measurements were only made with *Acer pensylvanicum* and *Populus grandidentata* because these are relatively large. Leaf-specific conductivities of last year's twigs (i.e., those with two growth rings) were between 10 and 30, and those of leaf-bearing shoots and of petioles were between 5 and 10. The lowest values were measured at the petiolar junction, namely 1 to 3. Petiolar junctions will be discussed in more detail in the next section.

At the time of this writing, LSC measurements have been made in coniferous species by Tyree et al. (1983) (*Thuja occidentalis*) and by Ewers (personal communication) (*Tsuga canadensis*). Tsuga had also been tentatively explored earlier, but this was before the discovery of the junction bottleneck (Zimmermann 1978 b). The principle of hydraulic construction in conifers seems to be the same as in dicotyledons: high LSC in the stem, lower LSC in branches, and constricted junctions. Tyree et al. found a rather good correlation between stem diameter and LSC in *Thuja*, reminiscent of Bailey's finding of the correlation between stem diameter and tracheid size (see Chap. 5.2).

It will be of considerable interest to link hydraulic architecture types to the morphological tree-model classification of Hallé and Oldeman (Hallé et al. 1978), although one must realize that the morphology is an expression of all functional adaptations, not only the hydraulic one. It will also be of considerable interest to study developmental events such as the changes in hydraulic properties of a lateral as it assumes the function of the main stem in sympodial branching, etc.

Let us now see how LSC values are related to pressure gradients. According to Eq. (3.2), volume flow is proportional to conductance times pressure gradient. As we are always dealing, at a given time, with the same volume of water on its way up from roots to leaves, LSC times pressure gradient must be constant. For example, if LSC in a branch is half that of the stem, the pressure gradient must be twice as steep. In order to calculate pressure gradients in a tree from LSC values, we must assume a certain transpiration rate. A few transpiration measurements were made with small detached maple branches before the leaves were removed for weighing (Zimmermann 1978 a). In full sunlight, the highest transpiration rates were of the order of 400–600 $\mu l\ h^{-1}\ g^{-1}$ fresh weight of leaves, and in open shade this dropped to 100–300 μl. These were maximum values, because water was supplied under atmospheric pressure to the twigs from a pipet, while in nature, water in the xylem is normally under tension. But these maximum transpiration rates permit us to estimate pressure gradients within the tree for comparative purposes.

Figure 4.6 shows the tree of Fig. 4.3 (left) and calculated pressure gradients in comparison with the hydrostatic slope. The figure is based upon two assumptions.

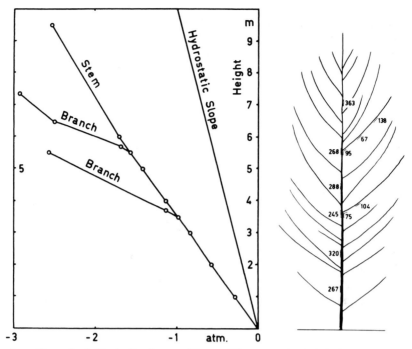

Fig. 4.6. Leaf-specific conductivities in *Populus grandidentata*. The tree on the *right* is the same as the one in Fig. 4.3 (*left*). The pressure gradients shown on the *left* have been calculated by assuming equal transpiration rates in all leaves. Pressure at ground level is assumed to be zero; if it were not zero, the scale would merely have to be shifted to the left or right. (Zimmermann 1982)

First, pressures at ground level are assumed to be zero. This does not have to be the case in reality, but it does not matter; a different pressure at that level would merely shift the pressure gradient to the left or to the right. In the absence of transpiration, pressure gradients would be identical with the hydrostatic slope. At maximum transpiration, all leaves transpiring at the same rate (the second assumption), the gradients would follow the curve drawn on the left. It can be seen that pressure gradients along branches are very much steeper than those of the trunk. This means that leaves on branches may experience lower pressures than leaves at the top of the tree, if they transpire at the same rate. It is very likely that their stomata close in response to the lowered pressure. The hydraulic design of the xylem path thus favors leaves at the top of the tree.

Measurements were made to investigate the anatomical basis of the characteristic LSC distribution within tree stems. Inside diameters were measured in 30–150 of the largest (i.e., hydraulically most significant) vessels at different points within the tree axes. Figure 4.7 shows the diameter distribution along the trunks of three birches and two poplars. The steady basipetal diameter increase, well known to plant anatomy previously (e.g., Fegel 1941) was thus illustrated. Vessel diameters in branches are distributed as in the main stem: they increase with increasing distance from the leaves.

The junction of branch and stem xylem is of particular interest. Junctions were perfused with dyes in basipetal direction. Different dyes were used for the main and

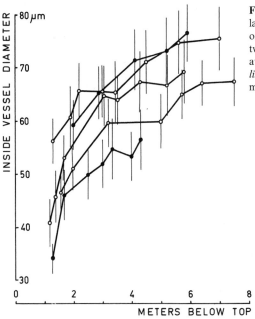

Fig. 4.7. Inside tangential diameters of the largest vessels in the most recent growth rings of the stems in three birches (*open circles*) and two poplars (*closed circles*). Each point is the average of 30–150 measurements. *Vertical lines* indicate standard deviations. (Zimmermann 1978 a)

lateral in such a way that one could identify that part of the xylem of the main stem which leads into the lateral (Fig. 4.8, inset). This showed that, as one followed vessels from the branch down into the stem, vessel diameters suddenly widen, i.e., they become "stem vessels." This is developmentally rather interesting. Although a branch controls cambial activity in the stem immediately below it, it does not seem to control vessel diameter.

But the most interesting aspect of a branch junction is its bottleneck. In the trees that were analyzed, the bottleneck was not due to a local drop of the Huber value, in fact the opposite was often found: a slight bulge in the basal part of the branch just outside the attachment. This may have been due to cambial stimulation at the point of greatest mechanical stress (comparable to the thickening of the trunk base). The constriction can have one or more of several causes. First, we often found a sharp drop in vessel diameter just above the branch junction (Fig. 4.8). This coincided approximately with the slight bulge. A second observation, made often in birch, was that many branch vessels just proximal to the branch attachment were non-conducting, often occluded by gums. This could have been caused by branches swaying in the wind. Vessels may have been torn and become momentarily leaky at the point of branch attachment; the reader should remember that a very minute leak is sufficient to cause cavitation. Vapor blockage is very often followed by gum secretion from paratracheal parenchyma (Chap. 6.3). A third possible cause of the bottleneck was found a few years later when vessel-length distribution was studied. In many species there are more vessel endings at a branch junction than along clear lengths of the axes.

Finally, let us see how LSC values relate to flow velocities. We might wonder, for example, whether flow velocity increases or decreases as the xylem sap moves

Fig. 4.8. Inside diameters of the largest vessels in the most recent growth rings of a poplar stem (*open circles*) (same tree as in Fig. 4.7, bottom curve). In addition, two branches are shown (*closed circles*), attached at 3 and 4 m below the top. Vessels of the branches are distinctly narower just a few centimeters above their attachment. Branch vessels inside the stem (identified by dye infusions) very quickly reached the diameter of the stem's own xylem. Standard deviations are comparable at all points. They are shown only at two points for the sake of clarity. *Inset drawing* indicates how dye perfusions with different colors identified stem and branch xylem. (Zimmermann 1978 a)

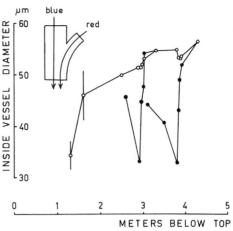

Fig. 4.9. The relationship of peak flow velocity, LSC and largest vessel diameters in different parts of a single plant, at any given time. The graph lets us, for example, compare velocities in stem and branch respectively. If the LSC of the branch is one half of that of the stem (LSC/2) and the largest vessel diameters in the branch are 0.8 compared with those in the stem (= 1), peak velocity in the branch would be 1.24 greater in the branch than in the stem. The curves go through the origin, of course. However, as relative diameter approaches zero, transverse-sectional xylem area would approach infinity

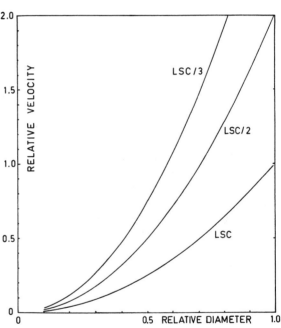

from main stem to branch. The answer is not a simple one because we should have comparative information on both the number of available vessels and their diameters. This information could be obtained by dye perfusions of the kind shown in Fig. 4.8 (inset), but it is cumbersome. A simpler method is to start with a comparison of relative pressure gradients (the reciprocal values of LSC). Next, we must measure the largest vessel diameters in the two axes that we want to compare. We measure the largest diameters only because we are concerned with peak velocities only. Velocities are directly proportional to the pressure gradient times the square of vessel diameter. The relationship is shown graphically in Fig. 4.9. Relative branch velocity, based upon stem velocity = 1, is plotted against relative vessel radius in the branch (stem vessel radius = 1). Three curves are given, one for equal LSC in the two axes that we compare, one for one half the LSC in the branch, and one for one third LSC in the branch. It can be seen from this graph that branch velocities can be either greater or smaller than stem velocities. Most commonly, LSC values are about half as great in branches as in the main stem. If, in such a case, branch vessels are more than 70% as wide as stem vessels, velocities in the branch are greater than in the stem [Huber and Schmidt's (1936) birch type]; if they are less than 70% as wide, velocities are smaller [Huber and Schmidt's (1936) oak type].

It is important that we realize that velocity itself tells us very little about the hydraulic construction of the plant. If we observe the transpiration stream with fluorescent tracer for example, and see how it moves faster in some parts and more slowly in others, and perhaps "hesitates" before entering the leaf petiole (Rouschal 1940), this does not mean much itself. It does not tell us anything about the magnitude of the pressure drop along the way and the pressure within an organ. Yet it is the pressure level which is important for the physiology of the plant; low pressure potential, for example, interferes with the proper metabolism of a tissue.

4.4 Leaf Insertions

The vascular connection between stem and leaf petiole is of particular interest, because the leaf is the most distal and most disposable part of the plant's segmented structure. The literature describing the anatomy of stem-leaf connections is very large. Howard (1974) gave a detailed survey of the variety of vascular connections between stem and leaves in dicotyledons. What interests us here is the question of how structural features affect the flow of ascending xylem sap. Meyer (1928), reviewing earlier literature and adding his own investigations, concluded that the xylem tracks of stem and petiole are rather separate. Stem vessels never continue into the petiole; leaf traces consist primarily of tracheids. He generalized this concept of "organ-restricted vascular bundles" so much that later investigators made some effort to prove him wrong. When he stated that no vessels are continuous from roots to stem, from stem to branch, and from shoot to petiole, he was probably correct as long as he dealt with the primary plant body, and possibly with very early secondary growth. We certainly know today that in secondary xylem of trees there are many vessels continuous from roots to stem, across branch junctions, etc. Rouschal (1940) studied water movement from stem to leaf with the aid of fluorescent dyes in relatively transparent herbaceous plants. His attention was concentrated on velocity and on transverse-sectional area. As we have seen in the previous section, these two features are not very informative by themselves. However, his method did yield interesting information. The dye moved up the main stem very fast and often entered top leaves before it entered bottom leaves. Rouschal also confirmed earlier reports of a xylem constriction at the petiole base (e.g., Salisbury 1913; Conway 1940).

Dimond (1966) conducted a very detailed study of pressure and flow relations in vascular bundles of the tomato plant. The tomato stem contains six vascular bundles along the stem, three large ones alternating with three small ones. One small bundle emerges from each node entering the petiole as a central bundle at the node above. Lateral petiolar bundles arise as branches from the two large bundles that lie on either side of the petiolar attachment. Flow rates were calculated from transpiration-rate measurements and from diameters of all individual vessels. These flow rates were then compared with experimental flow-rate measurements. What was not taken into account is the difficult question of flow resistance from vessel to vessel through intervessel pits. As we have seen in Chap. 1.3, this can be about as much as longitudinal resistance. Nevertheless, at the flow rates indicated by transpiration measurements, pressure gradients along the stem were of the order of 0.13 atm m^{-1}, along petioles of the order of 0.72 atm m^{-1}. Conductance of large (i.e., stem-) bundles decreased in the stem in apical direction, but fewer leaves have to be supplied higher up. The very highest leaves required less pressure drop to get water than intermediate leaves.

The paper by Begg and Turner (1970) also contains information about the leaf insertion. By making pressure measurements along tobacco stems on bagged and unbagged leaves, they found a maximum difference of 5.5 bar between stem and leaf. Other authors reported similar findings. But let us now return to the central problem of this section, namely the constructional principle which "separates" the leaf from the stem, so that the leaf may become disposable if the plant cannot afford to maintain it.

Larson and coworkers studied the development of the vascular system in shoots of *Populus deltoides* in very considerable detail. The poplar leaf petiole is supplied by three vascular bundles. By measuring individual vessel diameters along the way, Larson and Isebrands (1978) found at the base of the petiole a constricted zone where vessel diameters are narrower. If the relative conductance (the sum of the fourth powers of the vessel diameters) is recorded along the xylem path, there is a sharp drop at the constricted zone to about one fourth of the rest of the path. This constricted zone is located proximally to the abscission zone of the petiole (Isebrands and Larson 1977).

A constricted zone has also been detected in the form of an LSC drop across the petiole insertion of *Acer pensylvanicum* and *Populus grandidentata*, as mentioned in the previous section.

The significance of the hydraulic construction of the leaf insertion can be illustrated most dramatically with palms. Most palm species have a single perennial stem to which the disposable leaves are attached. This is hydraulically a very sharply defined case: the leaves are disposable because the stem cannot, under any circumstances, be lost. Furthermore, the stem xylem must remain functional for many years because the stem lacks a cambium. As one inspects the mature stem anatomically, one can see that leaf traces, containing narrow xylem elements, are inserted in (attached to) the vascular system of the stem which contains wide metaxylem vessels. This is the mature, functional construction with which we shall deal on the following pages. However, we must be aware of the fact that developmentally the leaf trace is not "attached" to the stem vascular system. A leaf trace is formed in direct distal continuation of the axial bundle. In other words, the axial bundle and its distal leaf trace is a single developmental unit to which the continuing axial bundle is attached later (Zimmermann and Tomlinson 1967; Zimmermann and Mattmuller 1982b). Still later, wide metaxylem vessels are formed along the axial

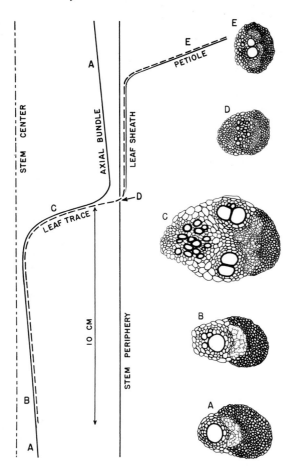

Fig. 4.10. Leaf-trace departure in the palm *Rhapis excelsa*. *Solid lines* indicate presence of metaxylem vessels. These are not necessarily continuous; a more precise plot of the vessel network is shown in Fig. 2.6. *Dashed lines* indicate narrow "protoxylem" tracheids. The letters *A* to *E* show transverse sections of vacular bundles and their location. An axial bundle (*A*) contains only metaxylem. Ten centimeters below the leaf-trace departure, the first evidence of protoxylem can be seen. At *B* a few protoxylem elements are present. *C* is a leaf trace just below the point where it breaks up into its branches. *D* is the leaf trace proper, the branch entering the petiole; it contains narrow tracheary elements only. *E* shows a vascular bundle in the petiole which contains again relatively wide metaxylem. (Zimmermann and Sperry 1983)

bundles, but not into the petiole. The axial bundle with its distal leaf trace is therefore a developmental unit to which the continuing axial bundle is attached. Functionally, however, the axial bundle system of the stem is a unit to which the leaf traces are attached. On the following pages we shall discuss this functional xylem system.

Figure 4.10 illustrates the principle of leaf trace attachment in the small palm *Rhapis excelsa*. This species is used as an example because we know it in most precise quantitative detail. It is, however, safe to state that the situation in other palm species is the same in principle (Zimmermann and Tomlinson 1974). The diagram on the left shows the course of a vascular bundle. The reader may want to refer back to Figs. 2.4–2.6 for general orientation. The solid line indicates the flow path provided by wide metaxylem vessels. This may not be a single, continuous vessel, but rather a succession of vessels with alternate tracks leading into neighboring axial bundles via bridges (see Fig. 2.6). The dashed line indicates the narrow-vessel path via leaf trace into the petiole. The drawings on the right illustrate transverse sections of a vascular bundle at different levels. It can be seen from this that the axial path along the stem is one of relatively large conductance through wide

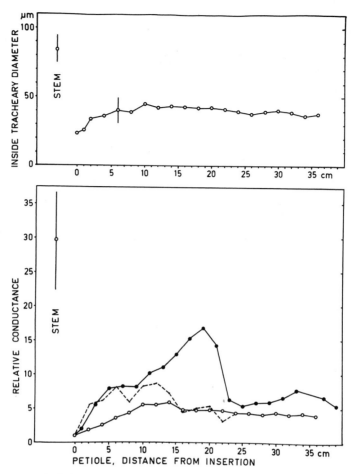

Fig. 4.11. Tracheary diameters and relative conductance along the stem (*point far left*) and the petiole from the insertion (including the leaf sheath) to the blade. *Above* average diameters of the largest tracheary elements; diameters are most narrow in the region of the leaf attachment (distance zero on the horizontal scale). *Below* relative conductance calculated as the sum of the fourth powers of all tracheary diameters, based upon the leaf insertion = 1. Three different petioles are shown. (Zimmermann and Sperry 1983)

metaxylem vessels. The path into the leaf is via narrow (protoxylem-like) tracheids. In the petiole itself there are again relatively wide metaxylem vessels. Dimensional measurements for the *Rhapis* leaf insertion are given in Fig. 4.11. Diameters are recorded in the upper graph. Diameters of stem metaxylem vessels are of the order of 85 μm, and in the petiole 45 μm. The narrowest diameters are found at the point of insertion itself. If we measure all inside diameters of the tracheary elements of the leaf trace leading into a petiole (about one hundred) and take the sum of the fourth powers, we obtain a measure of relative conductance. This is shown in Fig. 4.11 below, and is based upon the conductance = 1 at the insertion which represents the bottleneck. The relative conductance of the stem, calculated as one metaxylem vessel for every leaf-trace bundle, is about thirty times greater than that of the insertion. We have done this calculation with 27 other palm species and found ratios ranging from 1:5 to 1:267 (*Rhapis* 1:30).

The constriction indicated by summing the fourth powers of tracheary diameters does not indicate the full extent of the hydraulic constriction. First, the narrow tracheary elements of the insertion are mostly tracheids. The resistance to flow from one compartment into the next is probably greater in tracheids than in vessels. Second, and probably far more important, the flow resistance from the wide metaxylem vessels in the stem to the narrow leaf-trace tracheids is probably very considerable, although the total length of the contact area is ca. 10 cm in *Rhapis*. This is clearly illustrated in Fig. 2.5. The numbers 6, 4, and 3 correspond to the letters D, C, and B in Fig. 4.10. By inspecting these bundle transverse sections carefully, the reader will become aware of the fact that the contact between the wide metaxylem and the narrow leaf-trace tracheids is very poor indeed. There are very few vessel-to-tracheid pit areas; the two xylem strands are even separated by a layer of parenchyma in most places (e.g., in Fig. 2.5 at 4). It is extremely difficult to measure conductance from the stem metaxylem into the leaf-trace xylem (Zimmermann and Sperry 1983). Hydraulically then, the leaf is very sharply separated from the stem. The significance of this will be discussed in the next section.

4.5 The Significance of Plant Segmentation

We are only just beginning to appreciate the significance of the segmented structure of perennial higher plants. An arborescent monocotyledon, such as a palm, is perhaps the organism that demonstrates the situation most clearly. Let us consider a coconut palm that has a crown of approximately two dozen leaves. This number is more or less constant, new leaves are added at the top, and old leaves are shed at the base of the crown at a rate of about one leaf per month. This means that the visible crown renews itself within a period of about 2 years. There are an additional dozen or two leaf primordia hidden within the bud, but these do not concern us here. While producing leaves the palm also grows in height, but unlike dicotyledonous and coniferous trees, the vascular system of the stem, once formed, remains the same. There is no cambium that adds new vascular tissue periodically. The primary vascular tissue of the stem is overefficient when the palm is short, and it may become limiting as the palm approaches considerable height. It may well be possible that long-distance transport limits the height palms can ultimately attain, because, as the palm gains height, the same transpiration rate requires an ever-increasing pressure drop to lift the water to the top.

The palm stem thus represents many years of "investment." Leaves, on the other hand, are more or less disposable parts. From roots, via stem, to the leaves, water must move along a pressure gradient. Leaves are the organs from which water is lost. The plant must therefore be designed in such a way that if water loss is excessive, transpiration is shut down. This is, of course, the function of the stomata. But there is always a residual cuticular transpiration. As the pressure drops in the xylem because of continued water loss, water columns must be induced to break before irreparable damage is done to living cells. The plant accomplishes this by providing xylem ducts with pores of such size as to admit an "air seed" at a given negative pressure, thus vapor blocking the compartment that has been "air seeded" (Fig. 3.6). The problem with this mechanism is that it is irreversible unless the plant experiences positive xylem pressure soon thereafter, for example in herbs that produce positive root pressure (and guttate) at night (Chap. 3.4). One of the most important design requirements is that such vapor blockage does *not* happen in the stem. Leaves are renewable and can be dropped prematurely. The sharp conductance drop at the petiole insertion provides the structural basis for this: it lowers the pressure in the leaf xylem quite drastically as soon as water moves, to a value distinctly lower than that in the stem xylem. Some of the xylem ducts may thus be vapor-blocked, decreasing conductance further. The sharp conductance drop at the petiole insertion therefore provides the mechanism to sacrifice the leaf in order to save the stem, if conditions require such a drastic measure. The mechanism is perhaps comparable to that of a lizard who loses its tail if caught by it: the tail is sacrificed, thereby saving the lizard's body that can, in time, grow another tail.

Some palms drop their leaves by abscission while they are still green and healthy-looking, e.g., arecoid palms with a crown shaft like *Roystonea* when grown under ideal (humid) conditions. In other palms the oldest leaves merely dry out and remain attached to the trunk. If we take the conductance into a healthy leaf to be 100%, it may well deteriorate during the course of its life span until it is finally zero when it is completely dry. This gradual loss of conductance must follow a positive feedback mechanism (a vicious circle); the more vessels are embolized, the greater the pressure drop, hence still more vessels will fail. If this is true, then we should see very few leaves in the crown of a plant that grows under dry conditions or, for some reason, has lost many of its roots. At the other extreme are palms grown in a greenhouse under extremely favorable conditions of water supply. Old leaves remain green and functional for a long time.

Another feature of monocotyledonous leaf attachment is its meristematic leaf base. When a palm leaf unfolds and becomes functional, its base may still be meristematic and the xylem in the basal part may not yet have reached its full water-conducting capacity. This has been reported by Rouschal (1941) for grasses. Thus, the conductance of the attachment may, during the lifespan of the leaf, initially increase by maturation of the basal xylem, then decrease again by irreversible xylem failure. Within grass leaves Rouschal (1941) found very rapid long-distance water transport in approximately every other one of the parallel veins, and slow cross transfer of water into the veins located between the fast conductors. The difference between the fast and the slow conductors could not be seen by superficial observation, in transverse section, however, the greater vessel diameter of the fast conductors could be seen and measured. Rouschal also found that the xylem tracks of the leaf blade became fully connected to the stem only after leaf-base maturation. We do not yet fully understand all aspects of xylem-water supply to monocotyledonous leaves and have here a wide-open field of potentially very interesting future research.

Let us now look again at the hydraulic construction of dicotyledonous trees. As far as the leaf insertion is concerned, the situation is similar to that of the monocotyledons. It is perhaps not as extreme, but has the same effect. Excessively low pressures due to drought or loss of part of the root system will wilt leaves before vessels in the stem are lost by cavitation. But the segmentation of dicotyledonous trees is more extensive than that of palms. The main stem represents the greatest investment of the tree and should not, under any circumstances, be lost. Indeed, one can often find that in trees whose root system has suffered excessive damage, individual lateral branches die. It is possible that this loss is due, in part, to vapor blockage of the branch xylem. Again we are dealing here with a feedback mechanism. If the water supply to a lateral branch becomes poorer, it cannot support as many leaves; as the carbohydrate supply from leaves declines, cambial activity diminishes and decreases xylem development.

It would be interesting to study the hydraulic construction of fossil trees. I assume that a dichotomously branching tree was constructed according to the pipe model. At least successive daughter axes are equivalent. Adverse conditions thus could damage both stems equally. It has been suggested by P.B. Tomlinson (personal communication) that the hydraulic construction of the main stem with lateral branches was such a distinct adaptive advantage that it caused dichotomously branching trees to become extinct.

A few years ago Tomlinson (1978) described the morphology of New Zealand divaricating shrubs. These are characterized by microphylly, absence of terminal flowers, and interlacing branches of different orders, often so intricately that it is impossible to extricate a severed branch from a shrub. Interlacing can be caused by wide-angle branching or by zig-zag growth. In other words, we are dealing here with a very highly segmented construction. The interesting thing is that some divaricates are juvenile forms of species in which the adult form is an upright tree. The description suggests a construction with very low LSC values in the juvenile condition which is maintained until an extensive root system has been established. Once this has been accomplished, a highly conductive main stem is being formed. This is, of course, wild speculation, and it will certainly be worth while to study the hydraulic construction of these remarkable plants.

A final example of plant segmentation is the peculiar structure of certain desert shrubs whose stems become lobed and finally split longitudinally as a result of the death of strips of the cambium (Fahn 1974; Jones and Lord 1982). If there is not sufficient water available, some of the aerial shoots may die and thereby leave enough water so the remaining ones can survive.

I would like to close this chapter by pointing out another area about which we know little, namely the question whether small plants and colonies of lower plants (e.g., mosses) are constructed according to the segmentation principle which is able to "save" certain (perhaps younger) parts at the expense of other (perhaps older) parts. I do not believe anything is known about this. Furthermore, we do know that there are plants with nodal barriers (e.g., Meyer 1928). It would be very interesting to explore these from the point of view of hydraulic segmentation.

Chapter 5
Other Functional Adaptations

The subject matter of this book is multi-dimensionally related in so many different ways that it is difficult to accommodate it logically in a one-dimensional, linearly proceeding text representation. The organization of the subject material into chapters and sections has therefore been somewhat arbitrary. Frequent cross references help to restore some of the multi-dimensionality. There are nevertheless topics which do not fit easily into the linear text sequence, some of these have been gathered in this chapter under the heading *Other Functional Adaptations*. The chapter has thus become rather a collection of odds and ends. Functional adaptation is a very attractive topic for botanists who are interested in xylem evolution, but it tempts many to walk on thin ice. Arber (1920) wrote: "One of the unfortunate results, which followed the publication of *The Origin of Species*, was the acutely teleological turn thus given to the thoughts of biologists. On the theory that every existing organ and structure either has, or has had in the past, a special adaptive purpose and "survival value," it readily becomes a recognized habit to draw deductions as to function from structure, without checking such deductions experimentally." Too many structural features have more than a single function, and too many functions depend on several factors, correlations become therefore easily unreliable and one begins to speculate wildly. But there is more to it: random mutations may produce structural features that are functionally rather unimportant. It is rather futile to try to interpret such features for their adaptive value (van Steenis 1969; Baas 1976). Nevertheless, we must learn a great deal more about wood function before we can begin to speculate about adaptation. I suppose we are all entitled to a certain amount of speculation, although I normally prefer to reduce questions to basic simplicity which makes them accessible to experimental tests. This chapter is the place where I have most often disregarded my principle.

5.1 Radial Water Movement in the Stem

Leaves of deciduous trees are attached to shoots that have been formed during the current growing season. In the case where we are dealing with low-efficiency, low-risk trees, axial water conduction in the stem goes through several growth rings. As water moves from the small absorbing roots into older roots, it must cross over into older growth rings in order to reach the leaves. Radial water movement is also required during the course of secondary growth. The onset of cambial activity proceeds quite slowly basipetally in the main stem of diffuse-porous trees and conifers. The water supply to the distal young growth layers must therefore be initially supplied through the basal older layers.

We have relatively little experimental evidence of this radial transport, although there are two structural features which are obviously serving the purpose of radial water transport, the vessel (and tracheid) network and the rays. If one analyzes the network carefully, one finds primarily a tangential spread of the water path within a growth ring which is experimentally evidenced by dye ascents (e.g., Fig. 2.3). However, there is also a very slight radial contact between successive layers of tracheids and vessels, so that over long distances water can move axially in centrifugal and centripetal direction within the growth ring. In conifers, bordered pits can occasionally be found on tangential tracheid walls throughout the growth ring. They are particularly conspicuous when they are situated at the growth-ring border. These have been known for well over 100 years and have been discussed repeatedly (e.g., Laming and ter Welle 1971). Bosshard (1976) called them growth-ring bridges. Similarly, one often finds intervessel pit fields on the growth-ring border of secondary dicotyledonous wood, whereby the latest late-wood vessels connect directly with the earliest earlywood vessels of the next growth ring (e.g., MacDougal et al. 1929, pp. 56–65). Braun (1959) measured this contact quantitatively in a poplar species and found that about 30% of the earlywood vessels at the growth-ring border have contact with the latewood of the older ring.

In addition to these axial water paths which connect inner and outer xylem layers over long distances, the rays provide direct radial contacts. Certain conifers are well known for their ray tracheids, and we can assume that in this case radial movement of water is relatively efficiently provided for. Ray vessels are very rare in dicotyledons, they have been described for a few species only (e.g., Chattaway 1948; van Vliet 1976; Botosso and Gomes 1982). Nevertheless, ray transport of water seems to be easily possible, even in the absence of special conducting cells. This becomes obvious as one observes dye movement in the xylem: water appears to spread radially very easily along rays, probably through the walls (Fig. 5.1, below). Volume flow must be very small through any one ray, but sufficient if one considers the large number of rays available along the length of the tree stem.

In the terminal 2 years of growth of a shoot one can experimentally show the path of water from last year's xylem of the 2-year-old twigs, into this year's xylem of this year's twig. One merely has to inject a dye in backward direction, i.e., basipetally from the current shoot. The dye will then move back into both growth rings of the 2-year-old shoot, and the path can be studied by cinematographic analysis. Somewhat surprisingly, but indeed quite logically, one finds only a very small amount of 2-year-old xylem at the tip of the 2-year-old shoot. This is quite logical because this xylem had to supply only the terminal leaves of last year's shoot. The path of water from the tip end of this 2-year-old xylem leads axially directly into this year's xylem of this year's shoot. This is very difficult to describe with words, although easy to see in a film. Figure 5.1 (above) may help to clarify the situation.

Although we know where the water flows, we have little quantitative information about volume flow of water through individual rings and from ring to ring. Radial conductivity measurements have been made many times (e.g., Huber 1956), but these do not take into account movement resulting from the radial deviation of the axial path, which is probably very important.

In arborescent monocotyledons without secondary growth, radial channels are provided by the metaxylem vessels of the more or less radially running leaf traces within the stem, and their continuation in the continuing axial bundle and via bridges (Fig. 2.4). However, in those monocotyledons which have secondary

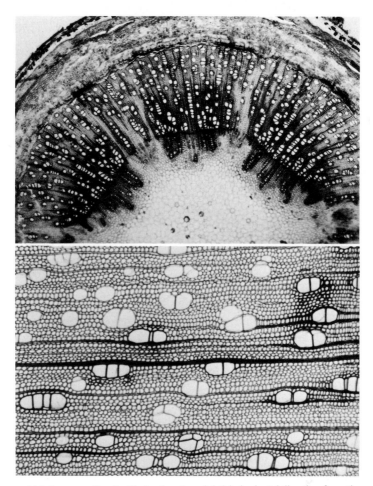

Fig. 5.1. *Above* A 2-year-old twig was perfused with dye (gentian violet) in basipetal direction from the current-year distal end, to show the path of water from last summer's growth ring of the 2-year-old twig into this year's ring of the current shoot to which the leaves are attached. This transverse section was made of the distal end of the 2-year-old axis. Heavy staining of the primary and early secondary xylem of last year's growth ring shows the direct axial path across the growth-ring border. *Below* A transverse section through the stem of *Acer rubrum*. A dye (gentian violet) has been pulled up into the wood through a few vessels (which can be recognized by their dark-appearing walls) and thus has been forced to move into other vessels. Dye has also moved radially through rays as indicated by their dark appearance. A color illustration would show the stain much more dramatically, of course.

xylem, radial paths are unknown and will have to be found experimentally, if they exist at all (Tomlinson and Zimmermann 1969; Zimmermann and Tomlinson 1970).

5.2 Xylem Structure in Different Parts of the Tree

There is a voluminous literature about the dimensions of xylem elements in different parts of trees. It begins with Nehemiah Grew (1641–1712) who noted that root-wood vessels are generally wider than those of the trunk (Baas 1982b). Systematic

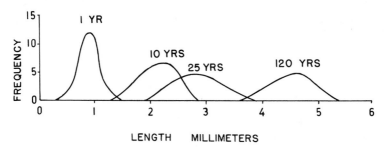

Fig. 5.2. Variations in length of tracheids with increasing age or diameter of a coniferous stem. In such a stem the fusiform initials of the cambium are of nearly equivalent length. Frequency distribution based upon measurement of 100 tracheids. (Redrawn from Bailey 1958)

investigations of this topic probably began with Sanio's (1872) classic work, describing his "five laws." This is not the place for a comprehensive review, our interest concerns specifically the relation of structure and function. However, it is very important that variations within an individual tree are respected. Variations of structure, not only within a species, but also those within an individual plant, have made, and still are making wood identifications difficult. For example, every wood anatomist knows that rootwood cannot be identified with a standard key, even branchwood may cause considerable difficulties. Paleobotanists had, and undoubtedly still have the same problem. Taxa of extinct plants are usually established on the basis of relatively small plant parts. It therefore happened that different genera were described on the basis of wood fragments from different parts of the same species (Bailey 1953).

Variations of anatomical characteristics from roots to twigs in an individual tree make it extremely dangerous to set up correlations of certain xylem features (such as vessel diameter) with habitat and draw conclusions about functional adaptations when one has only a random sample of wood from each species. Wood anatomy varies so much throughout a specimen that ecological trends based upon vessel diameters, densities, etc. make sense only if the anatomy of the compared species is known precisely throughout the individual plants.

The distribution of tracheid dimensions in coniferous trees has been summarized briefly, but thoroughly by I.W. Bailey (1958). The situation can be illustrated with two of his graphs. Tracheids become longer and wider with increasing age (diameter) of a stem (Fig. 5.2). There is also a distinct increase in length and width in basipetal direction from twigs to branches, to stem and finally to roots (Fig. 5.3). The two statements are of course related, except that the large tracheid size in roots is not strictly related to age, and certainly not to root diameter. Rundel and Stecker (1977) measured tracheid dimensions and xylem pressures in young branches at different heights in a 90-m-tall *Sequoiadendron giganteum*. Pressures followed the hydrostatic slope quite closely, and tracheid diameters were linearly correlated with pressure. Tracheid length, on the other hand, correlated very poorly with either diameter or pressure.

In woody dicotyledons, vessel diameters increase in basipetal direction from the top down (Fig. 4.7), they are usually greatest in the roots. Although this was known

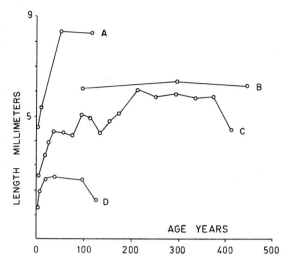

Fig. 5.3. Variations in average length of tracheids in stems and roots of *Sequoia sempervirens. A* Root 6.4 cm in diameter and 120 years old. *B* Root 30 cm in diameter and 450 years old. *C* Stem 1.5 m in diameter and 420 years old, at stump height. *D* Branch 4.55 cm in diameter and 130 years old, from the crown of a huge, old tree. Averages based upon measurements of 100 tracheids. (Redrawn from Bailey 1958)

to the earliest botanists, there are not nearly as many systematic investigations of dicotyledons as there are of conifers. The earliest systematic report known to me is a study of comparative anatomy of branch, trunk, and root wood of tree species of northeastern North America by Fegel (1941). Not only do vessel diameters increase basipetally from branches and down along the trunk, they often continue to increase in the roots with increasing distance from the trunk. This observation was reported by von Mohl (Riedl 1937), confirmed many times, and quantitatively represented by Fahn (1964, Fig. 3). But roots cannot always be included in such a statement, for example it is important that one distinguishes between the horizontal, rope-like conducting roots with their wide vessels, and the vertical taproots which serve primarily as storage organs (Riedl 1937).

Not only vessel diameters, but also vessel lengths increase in basipetal direction. Figure 5.4 shows the basipetal increase in vessel size of red maple. Even within the trunk there is a very distinct basipetal gradient of increasing vessel diameter and length (Zimmermann and Potter 1982, Fig. 2). At 11 m height, longest length was 20 cm and about 93% of the vessels were in the shortest length class (0–4 cm), and at the base of the trunk, longest length had increased to 30 cm and only 54% of the vessels were in the 0–4 cm length class.

Plant anatomists often measure vessel-element length in order to study "Baileyan trends" (Fig. 1.1). Tracheids of vessel-less woods are very long, length is obviously an hydraulic advantage. But as soon as vessels evolved, vessel elements shortened, because the hydraulic length requirement, previously filled by a single cell, was now filled by a cell series. In plants with vessels, cambial initial length is independent of hydraulic requirements.

I cannot believe in Carlquist's (1975) argument that shorter vessel elements are an adaptation to greater tensions because they make vessels mechanically stronger and thus prevent collapse. A little added wall thickness does this too well, and there are too many other factors involved. Vessel-element length, it seems to me, is merely a byproduct, e.g., the result of cambial initial length which might be under

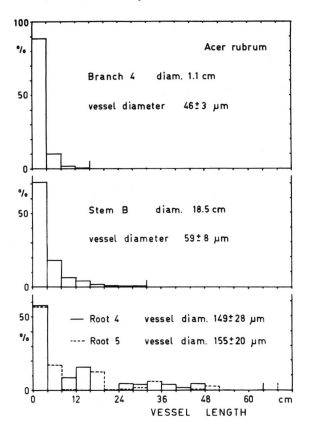

Fig. 5.4. Vessel dimensions in a branch, the main stem, and two roots of red maple (*Acer rubrum*). Vessel lengths and diameters (average and SD given) increase in basipetal direction. (Zimmermann and Potter 1982)

some other control, perhaps fiber length, although fibers can also elongate by intrusive growth. The variation of vessel-element length seems to me a perfect example of what Baas (1982b) calls "functionless trends imposed by correlative restraints."

Hydraulically then, vessel-less trees like conifers, as well as vessel-containing trees, are constructed in the same way: the conductance of individual compartments decreases acropetally from roots to leaves. The compartments are large and fewer in number at the base of the tree; they become smaller and more numerous in acropetal direction, i.e., they become safer (Chap. 1.4). This construction appears to be a perfect adaptation to prevailing pressure gradients. Pressures are always higher at the base of the tree and always decreasing acropetally, the danger of embolism therefore increases with height. Exceptions to this rule are probably very rare and restricted to brief periods after heavy rains. It is not surprising that the very largest vessel diameters (0.6–0.7 mm) found anywhere in the plant kingdom have been found in the conducting roots of wide-vessel species (Jeník 1978). Klotz (1978) found that in wide metaxylem elements of palms, the bars of the scalariform perforation plates are most closely spaced in leaves, intermediate in the stem, and farthest apart in the roots. The plates are also least oblique in the roots. From a functional point of view it is reasonable to propose that vessels originated in roots, stems, and leaves in succession during evolution (Cheadle 1953).

Vessels are narrow and short in seedlings. This has been shown dramatically in ring-porous trees. In American elm, for example, 1-m-tall seedlings had a longest vessel length of only 4 cm, and 91% of the vessels were in the shortest (0–1 cm) length class. In a 1.5-m-tall seedling, longest vessels length had increased to 26 cm and the percentage of vessels in the shortest length class (0–2 cm) had dropped to 70% (Newbanks et al. 1983). In mature elms, on the other hand, vessel lengths range over many meters. As a dicotyledonous tree grows in height, its water conducting system becomes more efficient from the base up. This is the same trend as the one found in conifers (Fig. 5.2).

Palms do not have secondary growth, in this case the xylem of the basal part of the stem must be made efficient enough for the mature plant from the beginning. The xylem of a short (young) palm stem is therefore over-efficient (Chap. 4.5). But the distribution of efficiency and safety features is shown here in another way. As we have seen in Chap. 2.2, the longest vessels are in the stem center, because here are the long (major) axial bundles. Vessels at the stem periphery are shorter, because here are the shorter (minor) axial bundles and the vast network of bridges. The peripheral xylem near the stem surface is therefore a safe network of vessels with many alternate pathways that can lead around possible injuries.

Finally, let us remember vessel distribution within individual growth rings at any one height. We have already seen that vessel width and length in dicotyledonous stems of temperate regions show a consistent decrease from early- to latewood (Fig. 1.8). This can perhaps be regarded as a built-in safety feature in temperate angiosperms, just as in present-day temperate conifers there is a certain specialization of early- and latewood. Earlywood is primarily conducting xylem, and latewood serves more mechanical function (e.g., Bailey 1958; Bosshard 1976).

5.3 Wall Sculptures and Scalariform Perforation Plates

We have seen in the very beginning of this book (Chap. 1.1) that the presence of scalariform perforation plates indicates an intermediate stage of vessel evolution. In many cases the intermediate character is quite obvious. In palms one finds all stages of transition from protoxylem tracheids to the wide metaxylem vessels which still have scalariform perforation plates (Klotz 1978; Zimmermann and Sperry 1983). There are other cases, for example species of the genus *Betula*, where the xylem consistently maintained the scalariform perforation plates (Fig. 5.5, lower right). It looks then as though the adaptive advantage of keeping them is greater than the advantage, namely decrease of flow resistance, of getting rid of them. What could be the nature of this advantage?

The function of scalariform perforation plates may be to catch bubbles when the ice in the vessels thaws at the end of the winter. It may have this function in common with other wall irregularities such as spiral thickenings (illustrated, for example, by Meylan and Butterfield 1978b). When Bailey (1933) investigated vestured pits (Chap. 1.5) he occasionally found vestures extending from the inside of the pit border to the inside of the vessel wall. These wall outgrowths have been studied in some detail by electron microscopy and are commonly called warts

(Frey-Wyssling et al. 1955; Liese and Ledbetter 1963; Côté and Day 1962). As will be discussed in the next chapter (Chap. 6.2), freezing displaces dissolved air and produces bubbles. These bubbles are present in the freshly thawed xylem water and present a grave danger. They must be dissolved before tensions appear in the xylem. The smaller the bubbles, the more easily and therefore faster they will dissolve. Plants that experience winters with freezing temperatures are in danger of suffering vapor blockage when such bubbles expand because of dropping xylem pressure when transpiration begins. One of the greatest dangers during the time of thawing is the coalescence of bubbles. Simple arithmetic tells us that eight coalescing spherical bubbles will form a single one twice the diameter. This not only drops the tensile strength of water to one half (Chap. 3.2), but it also will take a very much longer time for the bubble to dissolve. Not only does this single bubble have only half the surface area of the eight small ones, more important, the air is concentrated in one spot. It is therefore a very great advantage to keep bubbles separated upon thawing.

A few years ago I observed, with a horizontally mounted microscope, the melting of ice in thick radial sections of birch. As soon as bubbles were liberated from ice they began to rise. However, the slightest obstacle in their path caught them; the scalariform perforation plate was a particularly effective trap. These experiments have since then been expanded by Frank Ewers in my laboratory. In model experiments he froze air-saturated water in glass capillaries of different diameter. He then let them thaw in horizontal position and examined them periodically with the microscope. Even when the capillaries remained in horizontal position, the bubbles in larger capillaries fused and did not redissolve after several days when the smaller bubbles in the narrow capillaries already had dissolved. In additional experiments, Ewers (personal communication) froze pieces of wood which had been perfused with air-saturated water. He observed the upper transverse-sectional surface of the wood during thawing with the piece in an upright position. Observation was done with water immersion objectives and epi-illumination. In our local diffuse-porous species (of northern New England) no bubbles could be seen to emerge at the cut surface, while in ring-porous (i.e., large-vessel) species like ash large, obviously coalesced, bubbles appeared at the upper surface.

It makes no sense to argue that certain habitats require higher flow rates than others and thereby exert a selection pressure that eliminates scalariform perforation plates. Flow rates depend upon too many factors other than perforation plates (vessel width and length, effectiveness of intervessel pit, number of functioning vessels per leaf surface area, transpiration rates, etc.) to permit such a simplistic assumption. However, Baas (1976) reported that the percentage of genera with scalariform perforation plates increases with altitude and latitude. This would support the notion that scalariform perforation plates have a function (such as bubble catching) in cold climates. A similar trend was found for vessels with spiral thickenings (Fig. 5.5, left and upper right) (van der Graaff and Baas 1974; Carlquist 1975) which might well have a similar function. The thin spiral thickenings in some secondary woods are also very effective in strengthening vessel walls without increasing flow resistance excessively (Jeje and Zimmermann 1979).

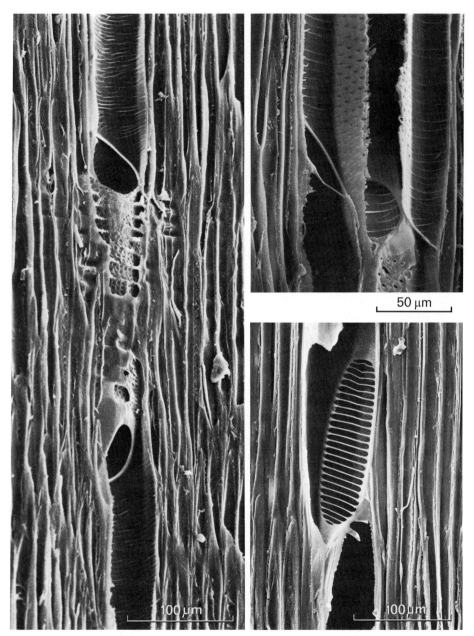

Fig. 5.5. *Left* Scanning electron micrograph of a longitudinal cut through secondary wood of red maple (*Acer rubrum*). A vessel has been cut in such a way that the central element remained intact but the elements above and below have been cut open. *Upper right* Scanning electron micrograph of a longitudinal cut of red maple wood, showing the spiral thickenings inside the vessel walls. *Lower right* Scanning electron micrograph showing the scalariform perforation plate between two vessel elements in birch. (*Betula* sp.)

5.4 Aquatic Angiosperms

Aquatic angiosperms are perhaps somewhat comparable to whales: they returned to the water, taking with them some features of terrestrial organisms. In whales, it appears to have been easier to modify the breathing system for a periodic return to the surface than to redevelop gills. In angiosperms, the root system had developed into such an important secretory organ in the sense of Kursanov (1957), that it was easier to maintain the xylem transport system than to have the shoot system regain functions that had been taken over by the roots. There is an enormous amount of literature about aquatic plants, but there are fortunately two quite thorough summaries, books which are not only informative, but make delightful reading as well (Arber 1920; Sculthorpe 1967). Furthermore, a reference volume on the anatomy of the Helobiae has just been published (Tomlinson 1982). We are here concerned only with questions related to xylem transport.

By 1861 Unger had already shown water transport from roots to shoots in two aquatic species (*Potamogeton crispus* and *Ranunculus fluitans*). The plants had their upper and lower parts in different containers while the central part was protected from drying by a bent glass tube. Unger recorded an increase of water volume in the upper container, but not when the roots were removed from the plant. The relatively high flow rates recorded by Unger could not be obtained by Wieler (1893), but Wieler demonstrated himself root pressure in many aquatics by observing bleeding after cutting off a shoot, raising the cut end just above the water level and enclosing the space above the cut end to maintain a humid atmosphere. Von Minden (1899) observed guttation from leaves of aquatics that were slightly raised above the water surface. When he wiped the water droplets from the water pores of the leaves, they quickly reappeared. Gardiner (1883) confirms Sachs' finding that root pressure depends on soil temperature. The phenomenon of guttation was already known at that time, Duchartre (1858) had discovered earlier that land plants, whose aerial parts were enclosed in a humid atmosphere, guttated water from leaves. By 1900 there still was a lively discussion in progress as to whether or not there was water transport in aquatic angiosperms; many prominent botanists were involved. As a further milestone cone can perhaps regard the paper by Snell (1908), a student of Goebel, who, in many careful experiments, established the roots as essential for nutrient uptake in many aquatics. When the roots were permitted to penetrate soil or sand, they developed root hairs. Even the floating plant *Pistia stratioides* (Araceae) needs the roots for proper growth. In two genera of the Lemnaceae, however, Snell (1908) found that roots were mere balancing organs, preventing the plants from flipping over. Here, water and mineral uptake was taking place primarily through the underside of the leaves. Snell (1908) had plants take up potassium ferrocyanide, which he could later detect in the vascular tissue by the Prussian blue reaction. The reported advance of ferrocyanide was of the order of 50 cm per day. It had thus been quite firmly established at the beginning of this century that roots and transport from roots to shoots are essential to growth of some aquatic angiosperms. The modern literature confirms this. Mantai and Newton (1982) found that the roots are very important for growth in *Myriophyllum spicatum*. They even reported that root growth was stimulated by a low mineral nutrient content of the water. Waisel et al. (1982) on the other hand found other species that are less dependent on the roots for mineral uptake. But there are also certain hormonal factors, such as cytokinins, which originate in the roots and must move via xylem to the shoot [see Bristow (1975) and the literature cited therein].

The question of the driving force of the xylem stream in aquatic plants is rather interesting; transpiration is out of the question in totally submerged plants. Root pressure is obviously the most likely candidate, it is suggested by the observation of bleeding from cut tops by Wieler (1893) and others. For example, Thut (1932) tested a number of species and showed water movement of around 10 µl h^{-1} per plant with roots; this figure dropped to one-tenth when the roots were removed. A driving force which could operate even in the absence of roots is a side effect of phloem transport. Münch's (1930) pressure-flow hypothesis, which is quite widely accepted today as the best explanation of phloem transport, requires that water be taken up osmotically into sieve tubes at the "source" and where photosynthesis takes place (i.e., in the leaves), and water be lost in "sinks," where translocated solutes are removed from the sieve tubes for storage or growth. This water returns to leaves via the xylem. Münch (1930) estimated it to be about 5% of the transpiration stream in land plants [for a discussion of this, see Zimmermann and Brown (1971), p. 263]. This component of xylem transport is independent of transpiration and could provide a significant portion of xylem water in aquatics. Thoday and Sykes (1909), working with submerged shoots of *Potamogeton lucens* in situ, carefully cut shoots and inserted them into vials containing eosin. The eosin solution was taken up and moved rapidly (up to 9.5 cm min^{-1}). When leaves were removed, movement of eosin dropped to 6 cm h^{-1}. As they were dealing with cut shoots, root pressure is out of question as a driving force, but ascent of water to the top due to sieve-tube loading would be a quite reasonable explanation of xylem transport in this case.

Thus, the literature leaves no doubt that xylem transport from roots to shoots is significant and essential in many aquatic angiosperms, although individual species differ in this. What concerns us most here is the structure of the xylem. The xylem of aquatic angiosperms is commonly considered to be reduced. But to call it "degenerate" is certainly misleading, because it is obvious that the structural requirements of a xylem that operates exclusively with positive pressures are fundamentally different from those of a xylem in which pressures are consistently, or periodically, negative. Xylem in which pressures are always positive should be constructed somewhat like resin ducts: it should consist of canals lined by living cells. Indeed, in perennially submerged plants we often find such xylem ducts (Fig. 5.6, right), while in plants whose tops are able to emerge from the water, "conventional," thick-walled xylem cells are present (Fig. 5.6, left).

While some of the fundamental questions concerning xylem transport have been reasonably answered, others have remained open and represent a fascinating area for further research:

First, what is the three-dimensional structure of the xylem? Have all compartment barriers disappeared, i.e., is the entire xylem system a single, complex cavity? This sounds very unreasonable, but we have little information about this point. There is some indication that these ducts are not continuous across nodal areas.

Second, and related to the first question is the problem of safety. If xylem translocates are of vital importance to the plant, how does the plant cope with injuries?

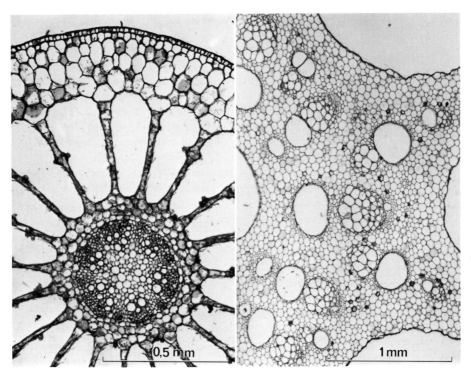

Fig. 5.6. *Left: Myriophyllum brasiliense,* transverse section of a submerged stem, showing large air ducts around the stele. The vascular tissue contains small, thick-walled tracheary elements. Emerging stems, when present, have a very similar structure. *Right: Nelumbo lutea,* transverse section of a stem. The large spaces at the upper and lower right and in the center on both sides, all partly cut off, are air ducts. Within the stem tissue are large vascular bundles with xylem and phloem. Most of these xylem ducts appear wall-less, but some have thin walls (e.g., the one at the *upper left*)

Xylem under positive pressure leaks when injured. The problem of sealing presents itself even in the absence of injury. There are plants that grow actively in front while they die away behind (Arber 1920). The xylem of the rear end must therefore be sealed off. What is this sealing mechanism? Rodger (1933) investigated wound healing in submerged plants. She found cell divisions (wound callus formation), cells growing into air ducts and thickening of cell walls. Sealing has also been observed when a gummy substance plugs the hydathodes of older leaves, thus directing the xylem stream to younger leaves (Wilson 1947). We must assume that cell walls are sealed off against injuries so that the xylem leak is plugged.

Third, intercellular spaces of land plants provide not only some water-storage space, but also air ducts (Chap. 3.4). In the case of frequent or perennial positive xylem pressures these ducts get flooded unless there is a special seal such as an endodermis or a bundle sheath that divides the apoplast into a water and an air compartment. In the case of aquatic plants the question then arises how the air-duct system, which is very well developed, is separated from the xylem. Where is this seal? How are the cells of the air-duct system supplied with root nutrients? In analogy to the situation in leaves and roots of land plants, one would expect xylem con-

tact via living cells and a seal (a Casparian strip) between the vascular tissue and
the air-duct system. On the other hand, we could assume that air pressure in the
air-duct system exceeds water pressure in the xylem apoplast. This might work as
long as the plant is uninjured. In case of injury to the air system however, the air
would escape and the air system would fill with water. Many aquatics subdivide
their air-duct system by water-repellent diaphragms which contain small pores that
permit passage of air but not water (see also Frey-Wyssling 1982).

Finally, there is the interesting case of similar xylem ducts in land plants; the
"carinal" canals in *Equisetum* (e.g., Bierhorst 1958) and the protoxylem lacunae in
monocotyledons. Dye movement experiments have shown that these canals can
transport water (e.g., Buchholz 1921). In land plants, the questions about function
are as follows: if these canals are continuous with the xylem, and there is hardly
doubt that this is the case, they could at least function as water-storage areas. The
first question which then arises is by what size pores they are connected to the intercellular air system. If the lacunae are tightly sealed off, the size of any air pocket
in them will have to follow the gas equation. In case of positive pressure the air
pocket is compressed and water storage space is thus provided. If the lacunae are
connected to the air system by larger pores, they should behave like (injured)
vessels. It is unlikely that lacunae can stand much tension without collapse, although in the case of *Equisetum* dye transport at below-atmospheric pressures has
been shown to occur in carinal canals (Bierhorst 1958). But the connections to the
air system could be dimensioned in such a way that air seeding takes place at pressures which do not yet cause collapse (Chap. 3.3).

Chapter 6
Failure and "Senescence" of Xylem Function

6.1 Embolism

In his review of xylem conduction, Huber (1956) wrote "we are virtually certain that the pressures of the moving water columns in the xylem are negative most of the time, but we are equally certain that such negative pressures cannot be maintained forever. Sooner or later water columns will break. Tracheary elements are thus filled with gas, probably irreversibly. This loss of water from the conducting channels is the beginning of a chain of events which we call heartwood formation." (Free translation from the German).

In conifers water loss from xylem is relatively easily recorded as water-content measurements. Figure 6.1 shows how the total wood volume of different growth rings is made up of cell wall material, bound water, free water, and "air." "Air" is written in quotation marks here because its quantitative composition may differ very considerably from atmospheric air. Free water, i.e., water in tracheids, decreases in successively older layers of wood as the number of embolized tracheids increases. Figure 6.1 represents a textbook example of water content, the heartwood is here relatively dry, i.e. most tracheids are embolized. Dry zones, some-

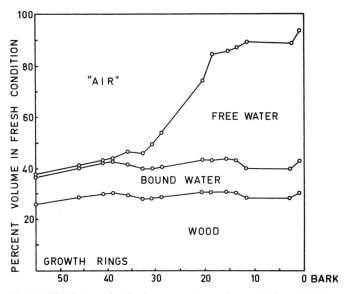

Fig. 6.1. Distribution of wall substance and water in growth rings of a 55-year-old spruce. The portion labeled *air* may be largely water-vapor space in embolized tracheids. Tracheids have been gradually embolized over a period of 40 years following their formation. Note that in this representation *distances between curves* (not from abscissa to curve) show amounts. (Trendelenburg 1939)

times described a pathological heartwood, can also appear in the sapwood as a result of fungal infection (Coutts 1976). This is undoubtedly due to premature embolism: fungal action makes tracheids leaky so that they embolize. On the other hand, heartwood has occasionally a rather high water content, it is then called wetwood. This special case will be discussed in Chap. 7.1.

Although vapor-blocked tracheary compartments are very difficult to demonstrate, our concept has not changed from that illustrated by Dixon (1914, his Figs. 17 and 18) (Fig. 3.13). Water transport through vessels of relatively transparent herbs has been observed directly with fluorescent dyes (Strugger 1940; Rouschal 1941). Direct observation of water transport is more difficult in woody plants. One normally uses dyes to mark the track, but in doing this the dye can also be drawn in by water-vapor-filled vessels (Preston 1952). This objection holds only if the dye is injected at atmospheric pressure, because the "vacuum"-filled vessels pull in dye that is administered at atmospheric pressure. Embolized vessels can be kept free of dye, if dye is applied under vacuum. We shall have to come back to this problem in Chap. 7.4. The vacuum method is difficult to use on entire trees. Dye ascent experiments stain cell walls only, not the tracheary lumen. One can often distinguish water-filled from vapor-blocked vessels by quick freezing. A cardboard box is mounted around a tree stem and filled with dry ice. When the stem is entirely frozen, it is cut with a bow saw right across the dry-ice-containing box, thus exposing a frozen transverse section which is kept cold, cleaned with a razor blade, and observed with a stereo microscope. Ice-filled vessels can be recognized, especially if they are large (Zimmermann 1965).

Embolism has been discussed a number of times in other sections of this book; we need not elaborate further upon it. However, it is perhaps good to emphasize that there are many different gas compartments in the xylem of a tree stem. Intercellular spaces provide the regular air-duct system (Chaps. 3.4 and 5.4). Parts of these may be tightly sealed from each other as discussed before. Tracheary compartments may be in contact with the air-duct system if they are badly injured. If they are merely air-seeded, they contain gas under a different, probably much lower pressure. Such gas spaces are isolated. There has been relatively little interest in the relation of water and gas space in woody stems, and this may well be a very fruitful area for research. For example, in many large palms the ground parenchyma of the (primary) stem expands considerably causing the stem to increase in diameter after the vascular tissue has matured (Schoute 1912) (Fig. 6.2). The parenchyma then forms a rather loose network of cells which criss-crosses what looks like an enormous volume of air space. It would be interesting to see if this space is required for respiration of the ground parenchyma, which, even in a very thick palm stem, must remain functional even in the center. Does it, under conditions of positive xylem pressure, get water-filled, or is it sealed from vascular tissues by bundle sheaths to prevent "suffocation?"

Our north-temperature ring-porous tree species (*Quercus, Fraxinus, Castanea, Ulmus*, etc.) are peculiar in that their earlywood vessels are so large that cavitation takes place at least by the end of the winter. The result is that most water is conducted in the wide earlywood vessels of the most recently formed growth ring. This means that these trees have to produce the earlywood vessels in the spring before the leaves emerge [Coster 1927; Huber 1935; Priestley et al. 1933, 1935].

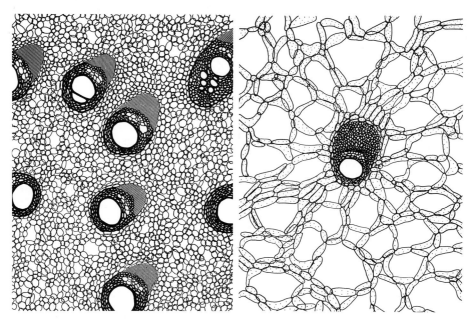

Fig. 6.2. Expanding tissue ("secondary thickening") in the stem of the royal palm. Suitable material from a single specimen was unfortunately not available, but would show about the same thing. *Left* Stem transverse section immediately below the crown. Ground parenchyma of the stem is still quite closely spaced, and the fibers of the vascular bundles are still extremely thin walled (*Roystonea elata*). *Right* Strem transverse section in the older and wider basal part of the stem. Ground parenchyma lacunose, fibers of vascular bundles thick-walled (*Roystonea regia*)

The wide vessels are so efficient that those of a single growth ring can supply the entire crown with water (Chap. 1.4). Ascent of sap during early spring while the new set of earlywood vessels is formed is via the small latewood vessels which remain functional for several years. Their capacity is sufficient to supply the leafless crown, but transpiration from leaves must await the formation of the wide earlywood vessels.

6.2 Winter Freezing

When water freezes, dissolved air is displaced and appears as bubbles. While the solubility of air in water is quite high (though temperature-dependent), it is at least 1,000 times lower in ice (Scholander et al. 1953). The question then arises of how trees of cold climates survive such massive air seeding in the spring.

The first question that must be answered is whether xylem water of trees does freeze in the winter. Any forester who has cut trees in the winter knows that the answer is yes. But the recording of the release of latent heat of fusion is certainly the most reliable detection of freezing. One occasionally encounters the myth that xylem water supercools substantially. This idea probably arose from the fact that it is difficult to freeze fast-moving water if the stem is chilled locally, because the chilled water continuously "escapes." Scholander et al. (1961), for example, were unable to freeze water in *Calamus* (a vine of the palm family) with dry ice unless they stopped the movement artificially. Many years ago I spent a summer studying low temperature effects in a number of North American tree species and found that supercooling amounted to no more than a few decrees C in all tested species (Zimmermann 1964). The freezing point of xylem water is only a few hundredths of a degree below $0°$ C, because the solute content of xylem water is rather low. While water in tracheary elements freezes rather easily, water outside the tracheary elements may supercool substantially (Burke et al. 1976; Hong and Sucoff 1982). Such "bound" water, some 30%–40% of the dry weight, may remain unfrozen (Lybeck 1959). (Note that the bound water shown in Fig. 6.1 amounts to ca. 30% if based on dry weight.)

The second question is whether xylem water contains dissolved air. The answer to this question is also yes. Scholander et al. (1955) found that the xylem sap of the grapevine is fully saturated with atmospheric nitrogen and partly with oxygen; there is no reason to believe that this is not the case in other plants.

Lybeck (1959) spun water in a Z-shaped capillary like Briggs (1950) (Chap. 3.2), at a speed which produced a tension of 1.8 atm in the center. When a piece of dry ice was brought into contact with the central part of the spinning capillary, the water column broke immediately. Scholander et al. (1961) studied the effect of freezing of xylem water in *Calamus*, the rattan vine, a genus of the palm family containing many species. The lower end of the vine had been cut and was taking up water from a burette. When a stem section was briefly frozen and thawed, water in the previously frozen section had evidently cavitated, because when the section was raised to a point 10 m above the burette, the water drained back to the burette instead of being transpired by the leaves. When the frozen-and-thawed section was lowered to ground level and the supply burette raised, full transpiration resumed as indicated by the rate of water uptake from the burette. It is quite obvious that this arrangement had refilled the embolized vessels with water by positive pressure (Fig. 6.3).

How do trees of cold climates cope with the problem of winter freezing? There seem to be at least three entirely different methods. First, there is the "throw-away" method. Ring-porous tree species have such large and therefore efficient vessels that a single growth increment of the trunk is sufficient to provide the crown with water. This has been discussed repeatedly in previous chapters. In this case, vessels are made before leaves emerge in the spring, a phenomenon that had been known to Coster (1927), and was discussed at some length by Priestley et al. (1933, 1935). The second method is the refilling method which is most clearly seen in the grapevine. Although the grapevine has rather large vessels (see Fig. 1.6) which embolize during the winter, they are refilled in the spring by positive root pressure as evidenced by the pressure gradients shown in Fig. 3.14. The third method is perhaps the most difficult one to understand. It seems that wood which contains tracheids only, or relatively small vessels, is not easily embolized by freezing.

Hammel (1967) froze 2–3-cm-long stem sections of twigs experimentally on standing trees with dry ice on summer days. He monitored the diurnal change of xylem pressure on nearby control twigs with the pressure chamber. Some time after freezing, he also measured pressures in the distal part of the frozen stem. In the four coniferous species he tested, freezing did not affect subsequent diurnal pressure

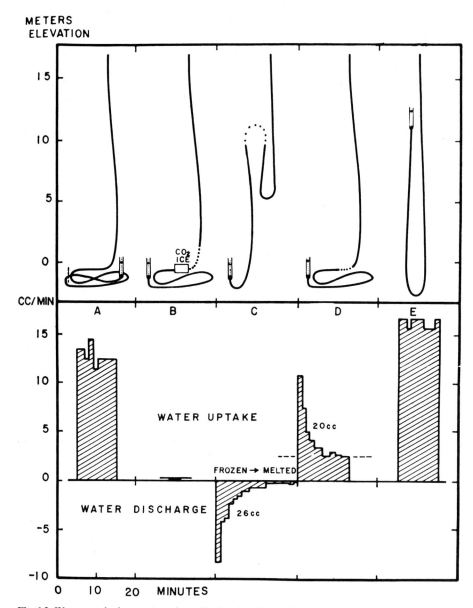

Fig. 6.3. Water uptake by a rattan vine: *A* before freezing; *B* frozen; *C* with nucleated loop elevated 11 m; *D* with cavitated section on ground; *E* with burette elevated 11 m. *Dotted line* cavitated section. (Scholander et al. 1961)

changes. In other words, flow resistance had not been increased substantially by freezing, i.e., much cavitation had not taken place. In dicotyledons, temporary freezing of the tensile water did increase flow resistance to greater or lesser extent, i.e., part of the frozen xylem path was inactivated by cavitation. When air-saturated bulk water freezes fast, air is trapped in the form of many small bubbles, when freezing is slower, fewer and larger bubbles appear (Carte 1961). Sucoff (1969) carried out experiments similar to those of Hammel. He estimated that ca. 9% of the water irreversibly migrated into the unfrozen part of the plant. Thus 9% of the water-filled tracheids would be lost to water conduction every time freezing occurs. This estimate is rather high as Fig. 6.1 indicates, especially if we consider that xylem may freeze and thaw repeatedly during the winter. The safety feature of smaller tracheary compartments is here again evident, the smaller the compartment, the more confined bubbles remain. To what extent, if at all, the valve action of the torus is involved in conifers, as Hammel (1967) suggests, we do not know. Both freezing and the displacement of dissolved air (bubble formation) cause a local volume increase. The process of freezing therefore brings about a local pressure increase. When thawing begins, this higher pressure may help dissolve bubbles. The smaller they are, the faster they will be dissolved. It is probable that the last compartments in which ice thaws are the most likely ones to suffer cavitation damage. Thus a certain percentage of the xylem may be lost each winter, the percentage being smaller the smaller the compartments. The chance of recovery is better the greater the effectiveness with which bubbles are prevented from coalescing within a vessel (Chap. 5.3). Readers who are specifically interested in this problem are referred to the papers cited above.

6.3 Tyloses, Gums, and Suberization

Tyloses had been seen and described by the very earliest microscopists. However, it was the Viennese baroness Hermine von Reichenbach (1845) who, in a remarkable paper, first described the origin of tyloses and who gave these structures their name. Her study was such a thorough and careful one that it makes many of the later papers about tyloses look redundant. Although she published anonymously, she was remembered for many decades. Nevertheless, she finally was forgotten. The rediscovery of her identity and her splendid work involved some interesting detective work (Zimmermann 1979).

Reichenbach (1845) described tyloses in many species, including some tropical plants. An English summary of her paper is given in Zimmermann (1979) where her 24 original drawings are also reproduced. She discovered that tyloses grow into the vessel lumen from neighboring parenchyma cells (Fig. 6.4 A, B). They are real cells, their nuclei can often be seen suspended by the cytoplasm; when neighboring cells touch each other, they form pit pairs (Fig. 6.4 C). Their cell walls often thicken secondarily, which makes the pit pairs more conspicuous. They often contain starch (Fig. 6.4 D). They behave in other ways like ordinary living cells, for example they can be plasmolyzed (Fig. 6.4 E) and the cytoplasm often shows streaming. Reichenbach's developmental studies are equally remarkable. In *Robinia pseudoacacia* she observed that all of the current year's vessels were free of tyloses and all older ones were filled with them. Tylosis formation in the current year's vessels began in October (in Austria) and was more or less completed in December. The exact same timing was reported recently for the same species from Kyoto, Japan by Fujita et al. (1978). Reichenbach (1845) concluded that tyloses form when the vessels are gas-filled, because they appeared in the fall after cessation of water con-

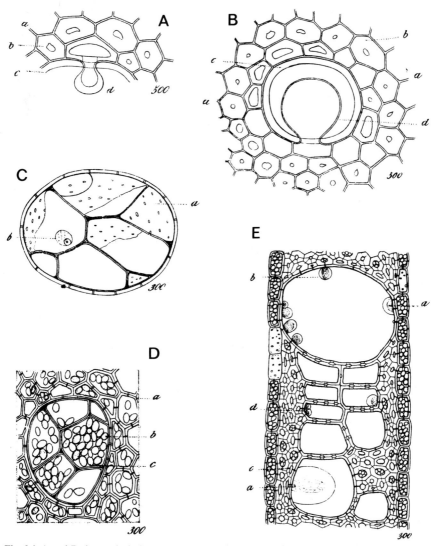

Fig. 6.4. A and **B** show tylosis formation in the stem of *Vitis vinifera*. The sections were treated with KOH: *a* primary cell wall; *b* secondary cell wall of wood parenchyma; *c* secondary vessel wall; *d* young tyloses with their mother cells. **C** Transverse section through a vessel of *Robinia pseudoacacia: a* walls of tyloses with pits in surface view; *b* nuclei. **D** Tyloses in *Vitis vinifera*, stained with iodine: *a* wood parenchyma cells; *b* tyloses containing starch; *c* vessel. **E** Vessels of *Robinia pseudoacacia* with young tyloses: *a* plasmolyzed cytoplasm. (Reichenbach 1845)

duction! They remained always in contact with their mother cells, because they could not get water or nutrients from the vessels.

Böhm (1867) showed that tyloses were produced regularly in connection with wounding. They always appeared in the living wood bordering dead wood, either near injuries or heartwood. He found that tyloses represent a very effective seal, they did not let water or air pass through vessels at his experimental pressures of

1–3 atm. Gum production had the same sealing effect in smaller cells. He therefore concluded that tyloses and gums provide the plant with a mechanism to seal off living tissue against injuries and other dead parts. Even Reichenbach (1845) had recognized the significance of gums: when she traced them in serial sections, she always found them to originate from injuries.

The sealing function of tyloses and gums was confirmed in many subsequent papers. When both these mechanisms were present in a single plant, tyloses were usually restricted to the larger, gums to the smaller cells (Praël 1888). There is much literature about tylosis formation in wood, in connection with leaf abscission, etc., which need not concern us any further here. Many of these papers indicated that tyloses and gums are the means by which vessels that had lost their water-conducting capacity were sealed off from functioning tissues. Tyloses are the only means in some plants, in others it is gums only, and in still others it is both. Chattaway (1949) found that in the many Australian species she investigated, parenchyma-to-vessel pits had to be at least 10 µm wide to permit tylosis formation. Furthermore, in her samples (from 1,100 genera!) tyloses and gums originated almost exclusively from ray cells.

The question what triggers tylosis formation was specifically addressed in a very detailed paper by Klein (1923). He repeated many of the earlier experiments of Wieler (1888) and others, and found that tyloses always appeared near wounds like drying cracks or cuts. His experiments were carried out with woody and herbaceous species, with stems, roots and aerial roots. Tyloses did not form as a response to the wound stimulus alone, nor as a response to the higher oxygen concentration of atmospheric air, nor to the cessation of water movement, but always as a response to loss of vessel water. Rarely, he found a few wide earlywood vessels in *Robinia* to be free of tyloses for a second summer, while the narrow latewood vessels were usually free of tyloses for several summers. Klein's (1923) findings remained available for many years while generations of plant physiology students repeated one of his experiments as a routine laboratory exercise. A Y-shaped piece of stem was turned upside down, one of its legs dipped into water, the other left in air, and suction applied at the morphological lower end. Tylosis formation could be observed in the half of the piece through which air had been drawn. The half through which water was drawn remained free of tyloses. Much of this older work was forgotten, and in the most recent literature we find the notion that tyloses may be plugging the vessels, thus preventing water transport, particularly as a result of infection. Such statements have no experimental support as far as I know (see Chap. 7.3).

Another misconception has crept into the recent literature, namely that tyloses are "ballooning" into vessels, implying pressure differences as a causal mechanism. Nothing could be farther from the truth. The pressure difference between the inside of a living parenchyma cell and the adjacent vessel lumen is greatest when the vessel is conducting, because the living cell is turgescent (i.e., has an internal positive pressure) while the vessel lumen may be under considerable negative pressure. When the vessel is embolized, the pressure in its lumen is between zero and 1 atm, while the cell wall surrounding the living cell still contains water under tension, i.e., the turgor remains the same. In other words, the pressure difference between the living parenchyma cell and the vessel lumen is very considerably greater while the vessel is functioning and tyloses do not form. We do not know how the absence of water on the vessel side stimulates the living parenchyma to grow into the vessel.

Electron microscopy made the study of ultrastructural details of tylosis formation possible. Readers interested in this are referred to the papers by Kórán and Côté (1965), Meyer and Côté (1968) and those of Shibata et al. (1981, earlier papers cited therein).

The gums mentioned above are produced by paratracheal parenchyma cells, primarily ray cells (Chattaway 1949), following stimulation such as cavitation or injury of vessels. But there are many species, coniferous as well as dicotyledons, that have special resin or gum ducts (Esau 1965; Fahn 1974). When stem tissue is injured, these resins or gums are released from the ducts, thereby soaking neighboring tissues and sealing them off effectively.

There is a growing literature on the physiological stimulation of gummosis. Readers interested in this topic may look at one of the recent papers and trace the literature from there (e.g., Olien and Bukovac 1982). The problem that is hardly ever addressed however, is the question of how these stimuli are linked to cavitation. If cavitation comes first, then conductance measurements are in vain (Chap. 7.4).

Tyloses and gums are means by which the plant seals off injured xylem which has become useless and could serve as entryways for infections. One or both of these mechanisms exist in many plants, but some plants have neither of them. But the injured tissue is not entirely isolated from the functioning tissue until the entire apoplast, including the cell walls, are sealed off. This is accomplished by suberization, lignification or other form of plugging of all cell walls that lead to the injury (Chap. 7.5). This seal is effective, like those of the endodermis in the root and leaf, but often broader in extent (not only one cell layer thick). It has been known for a long time and studied particularly in connection with abscission (Fahn 1974, p. 276; Addicott 1978). For plants that do not form tyloses or gums, it is the only seal. Some of these topics will have to be taken up again in the next chapter.

6.4 Heartwood Formation

The light outer sapwood is the living xylem, most recently produced by the cambium; it functions in water conduction and storage of starch and other substances. Tree trunks often have a dark core, the heartwood. This is the oldest, dead part of the wood. There are trees in which heartwood is always darker than sapwood and clearly recognizable; there are others in which there is a functional change, but no color change. In still others the color change is facultative. Finally there are species which show no color change and a very slow decline of physiological activity over time (Bosshard 1974). In some representatives of the first group with regular, darkly colored heartwood, there are tropical members in which the living xylem parenchyma cells survive for a very long time, so that the heartwood occupies a relatively thin core. Some very desirable types of wood, such as ebony (*Diospyros ebenum*) and letterwood (*Piratinera guianensis*) have this characteristic. This makes their desirable dark heartwood rather precious, because even a large tree yields relatively little. The opposite is usually the case in ring-porous species of the north-temperate regions where the sapwood layer may be rather thin. This again indicates that loss of vessel water is probably the first step in heartwood formation.

Heartwood formation is the normal consequence of xylem senescence. Measurable parameters of living processes, such as respiration, decline in the sapwood (Ziegler 1967). Between sap- and heartwood a transition zone can usually be recognized (Chattaway 1952; Shain and Mackay 1973). The dark coloration of "heartwood" is due to the deposition of lignins and polyphenols (Bauch et al. 1974; Hillis 1977) and may or may not involve centripetal transport in the rays, depending on species (Frey-Wyssling 1959; Höll 1975). Readers interested in heartwood formation are referred to the above references. What interests us here is its relation to water transport.

The percentage of xylem that remains functional over the years can be seen most easily in conifers (Fig. 6.1). The free water here normally represents mostly water in the lumen of the tracheids, i.e., movable water. Water content measurements of course also include extracellular water; this has been discussed in Chap. 3.4.

In species which have vessels the matter is more complex. This can be illustrated best with the example of the north-temperate ring-porous species. As previously discussed, these conduct water primarily in the large earlywood vessels of the most recent growth ring. The volume of this movable water is relatively small, because flow velocities are very high (Chap. 3.7). When water content measurements are made with pieces from successively deeper layers within ring-porous wood, the movable water is likely to be missed. The outermost layer, containing the most recent growth ring, should have the highest water content, but it usually does not. The water is very likely to be lost in sampling if the xylem is under stress when the tree is felled, and the water withdraws immediately from the cut surface. What is then recorded is (partly) the content of the latewood vessels, tracheids (if present), fibers, etc. The problem is, of course, the same in all vessel-containing wood types; it is less severe if there is little stress at the time of sampling, if the vessels are small, and if the samples are collected fast.

It has been said that the loss of water is the first step in heartwood formation. This is probably the reason why roots have in general very little heartwood and why proportionally more heartwood is often present higher up in the tree (Bosshard 1955; Ziegler 1968). Stresses are always least in the roots, greater higher up in the tree, and water loss by cavitation must therefore be greater at the top.

A final feature of interest in connection with heartwood formation is the status of the coniferous bordered pits. The discussion in Chap. 3.5 indicates that the coniferous bordered pits close when the pressure drop across them becomes excessive, i.e., when the water in one of the two contiguous tracheids cavitates. What we do not know is whether the pit membrane is flexible enough to regain its original position when the other tracheid also becomes vapor-filled, or if the membrane remains stuck permanently to the pit border (Comstock and Côté 1968). In other words, when we look at bordered pits with the microscope, we are not certain whether pit membranes in their normal central position (such as shown in Fig. 3.12, left) have always been in that position, i.e., that both tracheids have been water-filled at the time of sampling. One-sided embolism may have displaced the membrane, and later embolism of the neighboring tracheid may have permitted the membrane to regain its original central position. We might speculate that recovery is possible if the displacement is not maintained for a long time. In the sapwood–heartwood transition zone, displaced membranes are common. Unilateral embolism may have existed for a long time and the membrane may have finally been "glued" into position by incrustation with heartwood substances. The presence of incrusted membranes in their original, central position may suggest that this is the case. Krahmer and Côté (1963) reviewed much of the older literature about this matter and described and illustrated their own observations. Pit closure and incrustation of the pit membranes are of considerable practical importance, because they decrease the longitudinal permeability of coniferous heartwood quite drastically, thus making permeation of wood with preservative liquids difficult. The permeability of heartwood in dicotyledons is, of course, also much lower than that of the sapwood, because of tyloses and gums.

Chapter 7

Pathology of the Xylem

This chapter is probably the most difficult one for me to write, because in doing so, I have to step outside my field of expertise more often than I like. Yet, I feel that it is necessary to build a bridge between those of us who are concerned with xylem structure and the ascent of sap, and those plant pathologists who are interested in xylem dysfunction. I hope that my colleagues in plant pathology will not take offense for my "meddling in their affairs," but that instead this will be the beginning of a fruitful collaboration. The two groups can certainly learn a good deal from each other. I also hope that I shall hear from plant pathologists, in order to learn where my concepts are wrong and my knowledge is incomplete. This chapter is not a review of pathological disturbance of water relations in general. Such reviews are available (e.g., Ayres 1978). There are also many aspects of xylem dysfunction that are important after conduction has ceased; these are also outside the area of interest of this book. We are concerned here only with the ascent of sap and its disturbance.

7.1 Wetwood Formation

Figure 6.1 shows the "normal" distribution of the water content of coniferous wood of increasing age within a standing tree trunk. The water content is highest in the most recent growth ring and decreases toward the stem center. Such graphs are often somewhat irregular due to what we might call biological scatter. In addition, we discussed in Chap. 6.1 that in ring-porous trees we very likely fail to record the full water content of the most recent, conducting growth ring. The result is that graphs show only the scatter of non-functional wood and therefore look somewhat confusing.

In a normal tree, the heartwood is the driest part of the xylem because the lumen of most of the tracheary elements is vapor-blocked. However, there is a condition, common in some species, rare in others, whereby the water content of the heartwood is as great, or even greater than that of the sapwood. The heartwood is then referred to as wetwood. It may contain liquid under positive pressure while in the sapwood the transpiration stream moves along a gradient of negative pressures. Why is the water of the central wet core not drawn into the sapwood? In order to develop positive pressure, the heartwood must be sealed within the tree trunk. If this seal, whatever its origin, is semipermeable, and if the solute content of the central core is sufficiently great, the heartwood space could develop a positive pressure by drawing water by osmosis from the sapwood, until the osmotic equation is balanced. Another possibility is an impermeable seal, effective as a Casparian strip. In this case, the solute content of the wetwood would be unimportant, but there would have to be an internal source of water and of pressure. Water could be derived by microbial breakdown of wall substance and pressure could be produced by internal (microbial) gas production. Let us now turn from these theoretical speculations to reality. This will not be a thorough review of wetwood formation, readers interested in this are referred to the review by Hartley et al. (1961) and Murdoch's (1979) bibliographic list of publications about wetwood formation. Our only concern here will be the physiological problem outlined above.

Wetwood has been known to forestry for a long time, it occurs in a number of coniferous and dicotyledonous species and the mechanism of its formation does not have to be the same in every case. Wetwood water relations are not easy to understand, because most of the papers that have been published about it concern pathology, wood quality, effects on wood processing, etc. and are relatively unconcerned about physiology. Furthermore, I can certainly not claim to have read or even seen all the literature on this topic. Nevertheless, this is an attempt to explain how the high water content (and pressure) in the heartwood can be maintained next to the negative pressures in the sapwood.

Let us look first at American elm (*Ulmus americana*), where wetwood is very common. Wetwood becomes evident when it is exposed by a broken-off branch, frost crack, or by deep holes made for purposes such as fungicide injection. A liquid then emerges from the hole and may flow out for a long period of time, running down along the trunk. This liquid is often secondarily infected and may produce unsightly streaks along the trunk. If excess liquid is taken up by the sapwood, it moves up into the crown and may cause damage or even kill the tree.

The osmotic potential of wetwood liquid is quite considerable. Murdoch (1981) reported an average value of -14.7 atm (a range of -6.6 to -23.2 atm) for American elm, while xylem sap has an osmotic potential of only a fraction of an atm. He obtained an average value of -0.5 atm for sapwood "expressate;" the value of "expressate" must be greater than that of xylem sap, because it includes cell sap from parenchyma. At any rate, the osmotic potential difference between sap- and heartwood would permit a pressure build-up in the heartwood core, at least during periods of lesser xylem tensions, provided there is a semipermeable "membrane" between the two.

Wetwood often contains gas under pressure. Murdoch (1981) recorded pressures from zero (i.e., atmospheric) during the winter, to about 1 atm during the summer. Other authors reported pressures up to 4 atm (Carter 1945). While the gas content of sapwood is similar to that of the atmosphere, with possibly a somewhat lower oxygen and a much higher carbon dioxide content, wetwood may contain up to 50% methane (Hartley et al. 1961; Murdoch 1981). If the high water content of wetwood is maintained osmotically, gas pressure would simply have a cushioning effect. As osmosis in this system may work rather slowly, the gas pressure may reflect the daily average of the wetwood sap pressure. In fact, the entire system is so extensive that it may never reach real equilibrium.

The wetwood area of elm appears to act like a single, giant osmotic cell that is separated by a semipermeable "membrane" from the sapwood area. This can be visualized something like a Traube membrane, as early plant physiologists called it. In 1867 Traube described how semipermeable membranes can be prepared by precipitating a semipermeable substance, such as copper ferrocyanide. This can be done within the walls of a porous shell such as a clay container, thus providing a rigid osmotic cell that does not fail when pressure builds up. Much of the early work on osmosis was done with such cells. We can assume that the microorganisms of the wetwood cause the deposition of semipermeable material within the wall and intercellular spaces, thus surrounding the wetwood area with a Traube membrane. In American elm, Murdoch (1981) isolated and identified 14 species of bacteria and two species of yeasts from wetwood tissue. We can speculate that one or more of these produce the seal. Its chemical nature is not known, but the presence of a pressure difference between wetwood and sapwood makes it a necessity.

Wetwood in fir (*Abies*) appears to be at least superficially different. The liquid content of the heartwood core is not much greater than that of the sapwood, and the two are separated by a dry zone (Bauch 1973). However, the water content curve from sapwood to the adjacent dry zone looks quite normal, i.e., is comparable to that of sapwood and a normal transition zone, as shown in Fig. 6.1. But instead of remaining dry, the heartwood regains a high water content. The seal we are looking for must therefore be between the dry zone and the wetwood. Bauch (1973) made measurements of radial hydraulic conductance and found a decrease from sapwood to dry zone, and a drop to zero in the wetwood. Coutts and Rishbeth (1977) reported osmotic potentials of wetwood of -3 to -5 bar, and Worrall and Parmeter (1982) -5.3 to -7.6 bar. This is enough to keep the moisture content relatively high, but perhaps not always enough to have the trunk bleed when punctured. These authors also found that wetwood formation in fir did not have to involve microorganisms as it could be induced aseptically. The "seal" seems to be a product of the plant itself, it did not form below a girdle. Another rather peculiar finding of Coutts and Rishbeth (1977) was that latewood areas sealed off as wet areas first.

In conclusion we can say that the evidence supporting the osmotic concept is very strong. There must be a Traube membrane around the wetwood area and it must be permeable to water diffusion, otherwise wetwood could not keep bleeding for days or weeks. Solute content does make osmotic pressure possible and the system often appears to be cushioned by gas pressure. Permeability must be very poor, otherwise diurnal changes of wetwood sap (or gas) pressures would have to be recorded. It is rather ironic that a wound in the wetwood area which bleeds liquid for a long period of time thus appears to have the transpiration stream as a source of water, in spite of the fact that the pressure of the transpiration stream is negative most of the time!

7.2 Movement of Pathogens

In order to move in the xylem, a pathogen has to enter first. This usually happens through an injury. A rather unusual situation exists in the case of viruses. Although normally phloem-mobile, they can move freely through the xylem from vessel to vessel if their diameter is smaller than the intervessel pit membrane pores, ca. 25 nm depending upon plant species. If it is small enough, it can move through all cell walls with the apoplast water and thus reach the surface (plasmalemma) of any living cell; we shall come back to this later. Larger organisms have to dissolve enough of the cell wall to get in. The problem of entering a vessel containing water under tension is comparable to the problem of introducing an object into a container which is under positive pressure. Any injury of the container wall carries with it the danger of pressure loss. If a bacterium thus enters a vessel containing tensile water (water at negative pressure), it almost necessarily has to cause air seeding, whereby it is flushed either distally or basipetally to the end of the vessel with the retreating water. Subsequent entries are, of course, also possible; the conditions for these can be varied. In the case of large enough multiple injuries, rain water can enter broken vessels by capillarity.

Before further discussing entry of fungal hyphae into functioning vessels, let us consider how far a pathogen could proceed under the most "favorable" circumstances. The vessel-length calculations, explained in Chap. 1.2 (Figs. 1.3 and 1.4) are relevant here; the distance to which vessels are incapacitated upon injury is discussed also in Chap. 2.5. The number of injured vessels decreases very sharply as one moves away from an injury (Fig. 2.10). If a spore or bacterial suspension is inoculated via a fresh wound, the inoculant concentration diminishes with distance from the point of entry in a very similar way (e.g., Figs. 1–3 in Suhayda and Goodman 1981). Inoculant concentration also reflects three-dimensional xylem distribution in other ways, e.g., shows constrictions at nodes (Chap. 4.4) (Pomerleau and Mehran 1966). The maximum distance a pathogen moves as a result of inoculation depends on vessel length. One has to be fully aware of the vast differences in vessel-length distribution patterns in small shoots, seedlings or, at the other extreme, trunks of ring-porous trees (Chap. 5.2). It is interesting that elm varieties with narrower vessels are more resistant to Dutch elm disease (e.g., Elgersma 1970). We know now that narrower vessels are also shorter (Chap. 1.2). Smaller vessels limit the extent of inoculation.

The initial inoculation may be bidirectional from the point of injury. For example, fungal spores introduced into the vessels of a stem by the feeding activity of a beetle may be flushed up and down in the injured vessels. Downward movement, i.e., movement against the normal direction of the transpiration stream, must always be considered "inoculation movement" in damaged vessels, not movement in the undisturbed transpiration stream. When a vessel is opened up by an injury, the xylem water at the injury is exposed to ambient pressure ($+1$ atm). This means that there is now a pressure gradient away from the injury in both directions, up and down. This can flush the spores either up or down in the broken vessel until they reach the vessel end. The broken vessel is now permanently out of function and upward movement of water resumes around it.

The most important barriers to xylem movement of any particle after the initial inoculation are the intervessel pit membranes. These have been described in detail in Chap. 1.5 and their effectiveness is illustrated in Fig. 1.5 which shows paint particle penetration. The pores in intervessel pits are of the order of 25 nm; this is very small indeed (ca. 1/20 of the wavelength of visible light!). Once a pathogen has been flushed into an injured vessel, it can move passively in the transpiration stream only if it is small enough to get through intervessel pit membranes. Such movement must be strictly with the transpiration stream, i.e., under normal conditions in the general direction from roots to leaves, or into other transpiring organs such as flowers. Pores in intervessel pit membranes are so small that virus particles are the only candidates for unrestricted movement in the transpiration stream, and only the smaller types. Schneider and Worley (1959) published a report about the southern bean mosaic virus that can be interpreted in terms of movement across intervessel pits. They estimated the size of the virus particle to be of the order of 30 nm, i.e., very close to what we think is the size of the pores of intervessel pit membranes. They demonstrated xylem mobility by its movement across steam girdles (which blocks the phloem path). The interesting aspect of Schneider and Worley's paper in this context is the fact that they refer to a systemic host (*Phaseolus vulgaris* L. var. Black Valentine) and to a local lesion host (*P. vulgaris* L. var. Pinto). It would be interesting to see whether Black Valentine has slightly larger intervessel pit pores than Pinto, so that the virus is freely mobile in the xylem of the former, but not of the latter bean variety (see also Valverde and Fulton 1982).

Many of the pathogens that spread through the xylem, as fungal spores or bacteria, are very much larger than intervessel pit membrane pores. These organisms can only move from one vessel into the next by destructive action, e.g., by enzymatic dissolution of the vessel wall, probably the pit membrane, the weakest part of the wall. This kind of movement is not passive as in the case of virus particle movement described above, but it is an active penetration from vessel to vessel.

The interesting question now arises of whether some pathogens are able to enter vessels without causing cavitation, and can thus spread with the transpiration stream. If this is the case, it is a remarkable adaptation. We do know that certain insects ("sharpshooters") have accomplished the feat: some of them not only penetrate, with their mouth parts, vessels containing water under tension without cavitation, but seem to have no difficulties pumping water out (Mittler 1967). We must assume of course that the vessel water cavitates as soon as the insect withdraws its stylet bundle.

This brings us to the question whether fungal hyphae can penetrate vessel walls, produce conidia, and have the conidiospores float with the transpiration stream to the vessel ends, germinate and repeat the process, thus accomplishing distribution in the transpiration stream? This question is a very important one, and we do not know the answer. Any evidence I have seen is at best circumstantial. An important criterion has already been mentioned. Movement of spores with the transpiration stream *as well as against it* (i.e., downward in a tree), is evidence for a break of the water column at the point of entry. In Dutch elm disease, for example, upward movement of spores of *Ceratocystis ulmi* is faster than downward movement (Campana 1978). This might indicate not immediate, but somewhat delayed air seeding. It is known that the fungi penetrate cell walls often with microhyphae which dissolve the cell wall locally (e.g., Chou and Levi 1971), a mechanism that might accomplish a tight seal. The port of entry must remain sealed long enough to enable the fungus to produce conidiospores in order to have them released into the transpiration stream. It is likely that penetrating hyphae continue to dissolve cell wall material (after all, they have to do this in order to exist and grow), the moment of air seeding, if not immediate, could be somewhat delayed. It must be remembered that a crack of the order of 0.1 μm wide is sufficient to admit air under stress conditions. This is well below the resolution of the light microscope! To look for the first inoculation injury with the electron microscope is like looking for a needle in a haystack.

Electron microscopic inspection of wall penetration is interesting, but when looking at such pictures, we never know whether we see the 1st or the 50th penetration. During the first penetration the pit membrane may be under enormous physical stress, and tear if weakened. However, if the water column in the vessel has already been broken, the pressure difference is not very great, i.e., somewhere between zero and 0.977 atm (the difference between atmospheric and water-vapor pressure). Any hyphal penetration into a vessel whose water content has already cavitated could therefore look "intact." We must conclude from this brief discussion that fungal hyphae *might* penetrate functioning vessels, but evidence in support of this is is no more than circumstantial. Futhermore, it would be important to know how long a tight seal can be maintained after entry. We should expect considerable differences between species. It is very difficult to test for embolism experimentally; we shall have to come back to this problem in the section on flow resistance (Chap. 7.4).

Bacteria are very much larger than intervessel pit pores, and also larger than the pore size required to admit air under ordinary xylem tensions. It is practically unavoidable that the first bacterium penetrating a vessel must produce a hole in the vessel wall (pit membrane) large enough to induce cavitation. The entering bacterium would then be swept either up or down to the end of the vessel where it would lodge against the next vessel-to-vessel pit membrane, multiply in the humid space of the vapor-filled vessel, and eventually enter the next vessel destructively. *Xanthomonas campestris*, a bacterium causing black rot lesions in cabbage (*Brassica oleracea*), is an interesting example. It enters the hydrathodes and moves basipetally in the veins, i.e., against the normal xylem flow. It has been reported to plug xylem vessels (Sutton and Williams 1970), but plugging must be secondary, because the plugging material could not move against the transpiration stream in the xylem. In other words, the bacteria must have moved basipetally first by the gradient reversal caused by air seeding, gum plugging would then be secondary. In the case of herbaceous plants we must consider the possibility of repeated refilling of damaged vessels at night if xylem pressures are positive, and reemptying during the day, when pressures drop below atmospheric.

When multiplying, some bacteria form tight clumps (R.N. Goodman, personal communication). Therefore, if a bacterium is flushed against a pit membrane, it may multiply and plug the entire pit cavity before dissolving the pit membrane. If parts of this clump grow into the next vessel and finally detach, they could be carried in the transpiration stream without causing cavitation. This is the only mechanism by which I would visualize spread of bacteria in the intact transpiration stream. The matter obviously needs to be investigated experimentally.

As far as distribution is concerned, the question of whether embolism takes place at the time of entry may appear to be merely an academic one. However, it becomes more important when we discuss the cause of increased flow resistance, i.e., the cause of the plant's ultimate failure and death. Embolism at the time of pathogen entry puts the vessel out of function immediately, and any visible later blockage must be considered secondary (Chap. 7.4).

Finally, in order to move in the xylem, a pathogen has to be able to exist in xylem sap, i.e., in an osmotic potential of a fraction of an atmosphere. A protoplast containing a vacuole could possibly not contain a turgor pressure in the absence of a cell wall.

7.3 The Effect of Disease on Supply and Demand of Xylem Water

It should be quite clear by now that a drop in xylem pressure below a critical level causes cavitation and normally puts the xylem out of function permanently. The cause of such a pressure drop can be either a failing supply of water to the xylem by the roots, or excessive demand by transpiration.

Insufficient supply of water to the xylem, such as a prolonged drought, can cavitate xylem water, interrupt the water supply to leaves, and thereby kill the plant. The water supply to the stem can also be curtailed by failures of the root system, e.g., by damage caused by root parasites such as nematodes, etc.

Xylem pressure can also be dropped below the point of cavitation by excessive transpirational demand for water. The regulation of transpiration is rather complex, it is influenced directly by the water potential in the leaf (the hydropassive mechanism of regulation): stomata close when the water potential in the leaf drops below a certain level. But if the plant is sensitized, stomatal closure is induced hormonally (by abscisic acid), the plant can thus be drought-adapted (Raschke 1975). Abscisic acid can induce permanent closure and leaf senescence. It has been reported that other chemicals have the same effect (Thimann and Satler 1979). Senescence-inducing substances have to be present in mere hormonal concentrations to be effective. They do not have to destroy the leaf directly, but by closing stomata permanently, the leaf cannot carry on photosynthesis and is thus put out of function. Infection may cause a senescence-inducing substance to be produced at any point where it can enter the transpiration stream, be it along the stem, or in the roots. It may be a metabolite produced by a pathogen, or by the host plant under the influence of a pathogen. In this case the plant does not die from a failure of the xylem, but from lack of photosynthesis, although yellow, drooping leaves may simulate wilting. Lethal yellowing of coconut palms (and other palm species) is an example, although the death of the plant is not caused by leaf senescence alone, but also by phloem invasion by the pathogen. Lethal yellowing is a mycoplasma disease which appears to be transmitted to leaves by a leafhopper (Howard et al. 1979; Howard and Thomas 1980). The mycoplasma-like organisms multiply in the phloem and are transported to phloem sinks, such as roots, young leaves and inflorescences, and fruits. They can be found in great numbers in the sieve tubes of young inflorescences just proximal to necrotic tips (Parthasarathy 1974; McCoy 1978; Thomas 1979). Premature abscission of young fruits is the first visible symptom (phloem failure); at a later stage of the disease, mature leaves turn yellow and die. But yellowing is only a very late expression of leaf senescence. The beginning of senescence can be detected at least 2 weeks before fruits drop (the first visible symptom) by the consistently high xylem pressure in mature leaves, indicating stomatal closure (McDonough and Zimmermann 1979). In other words, even though the yellow leaves droop, lethal yellowing is not a wilt disease.

There are many diseases that result in yellowing leaves. It would be interesting to check diurnal xylem pressures in these and compare them with healthy controls, in order to see whether stomata are closed and thus the yellowing leaves are senescent. If this is the case, then xylem pressure measurements would indicate stomatal closure long before yellowing becomes obvious.

Under the headings *Shoot Diseases* and *Foliar Diseases*, Ayres (1978) reviews the effect of certain pathogen metabolites, such as fusicoccin, that open stomata. This causes transpiration in excess of available water and may not only cause wilting, but possibly also cavitation, i.e., permanent blockage of xylem.

7.4 Xylem Blockage

This section concerns xylem failure due to excessive flow resistance. It should not be regarded as a review of wilt diseases because there are many aspects of wilt diseases, such as permeability changes in living leaf cells, that have nothing to do with xylem structure and function. Readers are referred to the appropriate literature for this (e.g., Dimond 1970; Ayres 1978; Talboys 1978). The main purpose of this section is to point out an important aspect of xylem blockage which has received very little attention in the pathology literature: vapor blockage. In any healthy plant most tracheids and vessels eventually go out of function by cavitation, probably air seeding (see Chaps. 3.2, 3.3, 6.1). Water columns under tension are so vulnerable that cavitation is a likely result of any kind of disturbance, be it mechanical stress (for example by the force of wind), bubble formation by freezing, microbial wall degradation, etc. From a physiological point of view, cavitation is the first failure that has to be *assumed*. Penetration of a functional element that does not involve cavitation has to be regarded as a remarkable achievement by a microorganism, that must be *proved*.

The probability of cavitation during inoculation has already been discussed in Chap. 7.2. What has not yet been discussed is how the problem can be investigated. A vapor-blocked section of a plant stem has an increased flow resistance as beautifully explained by Scholander in a number of papers (e.g., Scholander 1958, Figs. 1–2 and 1–4; see also Fig. 6.3). In the case of Scholander's experiments, the pressure drop in the xylem distal to the blockage was produced by air intake at the cut end or by air seeding, caused by freezing. The question now arises how we can distinguish between cavitation and physical plugging caused by hyphae, bacteria, or some high molecular solute.

In experimental tests, the problem is very easily missed. First, wilting of cuttings whose xylem has been supplied with a solution of a "toxin" (e.g., Ipsen and Abul-Haji 1982) certainly does not give us an answer; just about any suspension or solution of a high-molecular substance will plug the xylem simply because the pores in the intervessel pit membranes are so small (ca. 25 nm!) and the investigator cuts the xylem first! Second, let us assume that the xylem path of a plant has suffered extensive embolization. As stem xylem is normally quite overefficient, wilting may not be immediately evident. If we now cut the stem and put the plant into a dye solution, the vapor-blocked vessels will refill as soon as the cut end of the shoot is brought into contact with the dye solution which is at atmospheric pressure. Neither the dye pattern nor the transpiration rate show the blockage. If we remove a piece of stem and make conductance measurements, the same thing happens: in contact with water at atmospheric pressure the xylem refills and conductance is back to normal. In other words, all these procedures do not show damage to xylem function that has taken place by cavitation. Dye movements and conductance measurements under ordinary atmospheric conditions in the laboratory are therefore of little value. When a conductance decrease finally does become evident, is very likely is the result of much later, secondary plugging.

In the late 19th century many botanists described transpiration experiments, whereby the cut end of a shoot had to take up water from a container in which air pressure was lowered by a vacuum pump. Air pressure can, of course, only be lowered to slightly above the vapor pressure of water (0.023 atm), otherwise the water would boil. But this is low enough to remove most of the water from vessels in which cavitation has taken place. Ursprung (1913) found that while even a clay wick continued to evaporate water with undiminished intensity when vacuum was applied to the container from which it was supplied with water, branches of *Robinia pseudoacacia* wilted. This situation is actually almost exactly identical with that of picking flowers in the garden or field, and putting them later into a vase. When one severs a flower (or a branch) from the intact plant, transpiration continues, thus withdrawing all water from the cut vessels into which, after some time, air is drawn through the cut end. From that time on, water has to come out of storage (Chap. 3.4), stomata close, and if one waits long enough before putting the flowers into a vase, wilting will begin. As soon as the cut ends are put into water, submersed cell walls take up water and pass it on to the nearest intact tracheids and vessels. If the capacity of this apoplast pathway to the nearest, distal intact xylem is sufficient, the flower (or branch) will not wilt. If it is not sufficient to support transpiration, the flower (or branch) wilts. The depth to which the cut end of the stem is immersed is very important (see Fig. 2.10). Most flowers survive the ordeal of picking and embolism at the cut end, but it is well known that some do not (*Anemone nemorosa* of the western European forests, branches of *Fraxinus* sp., etc.).

Transpiration from a vacuum container is an excellent tool to test for embolized vessels. Figure 7.1 shows the set-up that was used by the early botanists, but it has been elaborated in several ways to serve our purpose. First, in order to maintain a reasonable vacuum even in the presence of small leaks, it is desirable to run the vacuum pump periodically throughout the experiment. A timer turns the pump on at intervals. A valve between the system and the pump is opened after the pump is turned on and closed before the pump is turned off to avoid vacuum loss via the pump during its rest periods. Second, the vacuum system includes a lower chamber that permits drainage of the main chamber which contains the plant. Liquid can thus be changed (i.e., successive liquids can be offered to the plant) without release of the vacuum.

In order to test the effect of a disease on cavitation, the following procedure was adopted (Newbanks et al. 1983). A branch or seedling of the test plant was infected by puncturing the stem with a hypodermic syringe whose needle had had its tip cut off. A spore suspension was injected into the xylem; control stems received water. After successive periods of time, branches or seedlings were harvested, fitted tightly into the vacuum container and the pump was turned on. We must assume that soon after the vacuum is established, transpiration draws water from all vessels which are cut at the basal end of the branch, as well as from all vessels that were ruptured by the inoculation needle. The main chamber was then filled carefully from the funnel with solutions of a basic dye. The best results (i.e., the shar-

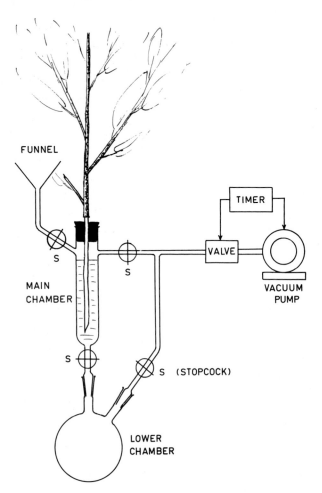

Fig. 7.1. Experimental arrangement to let a cutting transpire while taking up water (or a dye solution) from a container under vacuum (reduced pressure). Certain plants, such as ring-porous tree species require that the entire root system of a seedling be immersed. If the experiment lasts a long time, the vacuum pump is run intermittently; a timer closes the valve before the pump is shut off, and opens it after the pump has been turned on

pest dye tracks) were obtained with Schiff's reagent. The plant was first given periodic acid for an hour or two, the main chamber was then drained via the lower chamber and Schiff's reagent introduced via funnel into the main chamber, without ever losing the vacuum. After a period of uptake of Schiff's reagent, the main chamber was drained again without loss of vaccum and the plant left under vacuum until wilting. This precaution assured that embolized vessels are not refilled by water-vapor pressure when the vacuum is finally released. Serial sections were made of the stem xylem, and the vessel network reconstructed by cinematographic analysis (Chap. 2.1). As we look at the resulting vessel-network reconstruction we know that all vessels whose walls appear stained have been in contact with the moving xylem liquid during the experiment. If a cut shoot is used, the analyzed portion of the stem should be at least as far away from the cut end as the length of the longest vessel. Cut vessels are thus kept outside the analyzed area.

Figure 7.2 shows the reconstruction of part of the vessel network of an elm seedling which was used as a control plant. Elm, by the way, is a ring-porous tree

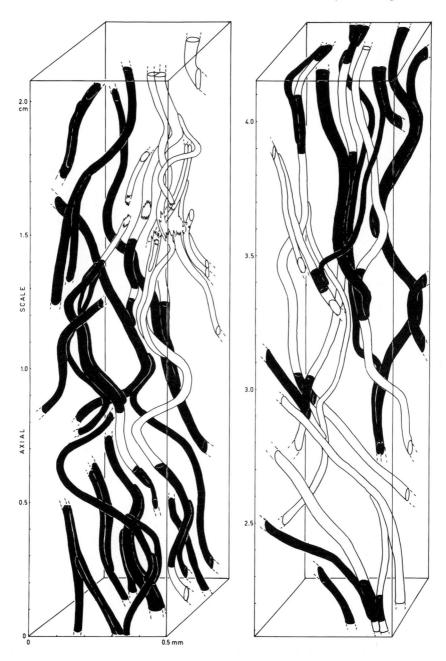

Fig. 7.2. Vessel network reconstruction of a section of a stem of a seedling of *Ulmus americana*. The block on the *right* belongs on top of the one on the *left*, as the axial scale indicates. This is a control stem which has been punctured and injected with distilled water. The puncture can be recognized by the torn vessels at 1.5 cm height of the axial scale. All functional vessels are dye stained throughout their entire length (shown *black*). Injured vessels were air-blocked and did not conduct dye solution. They appear unstained, expect where their walls were stained by contact with conducting vessels. Note that the axial scale is foreshortened ca. ten times

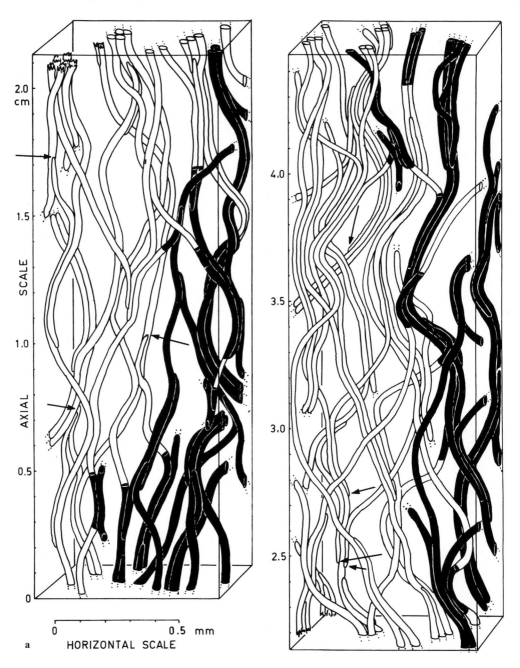

Fig. 7.3 a, b. Vessel network reconstruction of a section of a stem of a seedling of *Ulmus americana*, injected at 2.1 cm of the axial scale with a spore suspension of the Dutch elm disease fungus *Ceratocystis ulmi*, 4 days before harvest of the stem. Many of the vessels beyond the injection injury have already become non-conducting, they are marked by *arrows*. As absolutely no microscopic evidence, such as

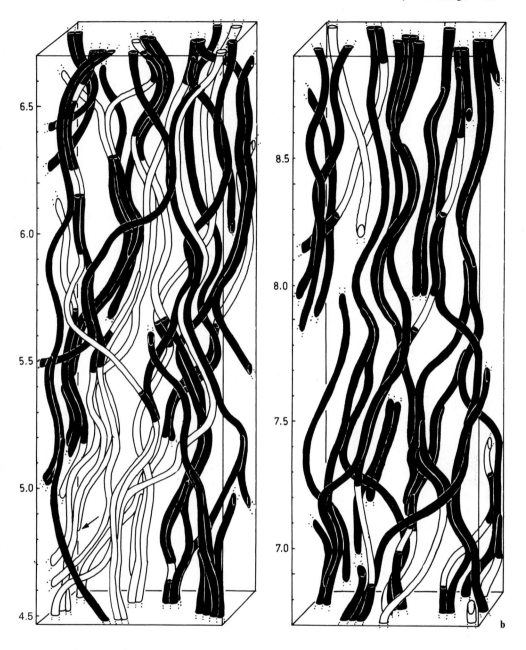

fungal hyphae, gums, etc., was visible, it must be assumed that the fungus damaged vessels walls enough to cause cavitation. The four pieces shown are all a single block of wood. From **a** to **b**, each piece on the right belongs on top of the piece on the left, as the axial scale indicates. Note that the axial scale is foreshortened about ten times with very poor water uptake from the cut end under vacuum. Sufficient water up-take was achieved by using seedlings and submersing the entire root system. The root system was probably killed by the contact with periodic acid, but this does not matter. Figure 7.2 shows a piece of stem that was 4.2 cm long and had a transverse-sectional area of 0.25 mm^2. The puncture of the needle can be recognized by torn vessels at 1.5 cm of the vertical scale. These "white" vessels were unstained. The vessels appearing black in the drawing were stained. It is important to realize that what we see in the microscope is only vessel walls, the method does not show the presence of liquid in the vessel lumen. Wall staining can take place not only by liq-uid flowing through the vessel lumen, but also by liquid seeping into the wall of an embolized vessel from a neighboring, conducting one. We refer to this as contact staining. Figure 7.2 shows that all embolized ("white") vessels are stained where they are in contact with dye-conducting vessels. In other words, any conducting vessel looks stained (black in the illustration) *throughout its entire length;* any vessel appearing partly white in the drawing must be regarded as an embolized vessel. If such an experiment is made without application of vacuum, results are much more ambiguous: dye is drawn into embolized vessels by the contraction of emboli and or by capillarity.

Figure 7.3 shows the reconstructed vessel network of an elm seedling that was infected with a spore suspension of the Dutch elm disease fungus *Ceratocystis ulmi.* The dye ascent was made under vacuum 4 days after inoculation. The recon-structed xylem block measured ca. $0.5 \times 0.8 \times 89$ mm. The inoculation puncture appears on the left side at about 2.1 cm of the axial scale, the torn vessels are visible at the top of the first and the bottom of the second block. All punctured vessels are again "white" except for the usual points of contact staining. But we can now see many more "white," i.e., embolized vessels, located entirely outside the area of physical injury. Upper ends of embolized vessels located below the injury and lower ends of embolized vessels above the injury are marked with arrows in Fig. 7.3. These vessels had been vapor-blocked, but not visibly injured. They showed ab-solutely no microscopic evidence of fungal activity (hyphae, gums, etc.), but they must have been submicroscopically leaky when the dye ascent was made, probably by enzymatic action of the fungus before fungal hyphae and gums become visible microscopically and before visible symptoms such as wilting develop. Many visible symptoms have been known for years, but the real first damage, vapor blockage, has received very little attention (VanAlfen and MacHardy 1978).

The above description of xylem dysfunction caused by the Dutch elm disease fungus is certainly not complete. Readers interested in a more comprehensive sum-mary are referred to recent reviews (Sinclair and Campana 1978; Stipes and Cam-pana 1981). The purpose of the present description is to stress the immediate and difficult-to-detect early damage caused by vapor blockage.

Another interesting case of damage by vapor blockage is the chestnut blight, the canker formation in the bark of chestnut trees (*Castanea* spp.) by the fungus *Endothia parasitica.* The fungus eventually girdles the tree stem, and thus kills the tree. Evidence of this is the sudden wilting of the entire part of the aerial stem distal to the girdle. There are also canker diseases of elm, butternut, and other ring-por-ous species that have similar effects. This situation may be confusing, because the immediate effect of girdling is not phloem- but xylem interruption.

Perennial shoots, particularly trees of advanced age, do not die immediately from the interruption of the phloem. The interruption of phloem kills only very young, phloem-importing shoots quickly. Every forester knows that diffuse-porous trees and conifers can survive girdling for many years. This has been documented many times. One of the famous cases is the Y-shaped, 100-year-old *Pinus sylvestris* mentioned earlier in this book. One half of it was girdled in 1871 for experimental purposes by Theodor Hartig (who had discovered the sieve tubes in 1837). Before he died, he asked his son Robert (who is known for the Hartig net of mycorrhizae and is often called the Father of Forest Pathology) to observe how long the girdled part would survive. It never came to this, because the tree had to be felled for other reasons during the winter of 1888–1889. The girdled part had survived for 18 years and still looked healthy, although its foliage was somewhat thinner than that of the ungirdled crown. Robert Hartig published a very detailed analysis of the growth distribution in 1889. Another famous case is a horsechestnut (*Aesculus hippocastanum*) in Belgium that survived complete girdling for over 50 years (Martens 1971). Girdling has also been used experimentally, in order to produce dwarf apple trees by the inversion of bark rings (Sax 1954). There are even examples of survival after pathological girdling: beeches (*Fagus grandifolia*) are often completely girdled by the fungus of the beech bark disease (a *Nectria* sp.) but they survive for a long time, often until the tree is blown over when, during a strong wind, the weakened trunk breaks. It can then be seen that the outer sapwood has been rotted by (secondary?) fungal infections, and thus weakened mechanically. The fact that the crown still contains healthy, green leaves when the tree falls down indicates that there was enough inner sapwood left to supply the crown with water.

Thus any immediate damage of girdling is almost always the result of xylem, not phloem interruption. It is evident that the trees most vulnerable to girdling are the ring-porous species which depend almost exclusively upon the most recent, superficially located earlywood vessels for their water supply. When a tree is mechanically girdled, these superficially located vessels are invariably damaged and the tree dies from xylem interruption. In the case of the chestnut blight, the fungus appears to make the earlywood vessels leaky either by enzymatic action or simply by drying when the bark dies. Although the circumstantial evidence in support of this view is very strong, direct experimental evidence is still missing.

It is interesting to look at xylem failure, particularly of ring-porous trees, again in view of Braun's (1963, 1970) *Hydrosystem*. Braun assumes xylem evolution in dicotyledons from wood with tracheids only, to wood with vessels within a matrix of tracheids and fibers (such as chestnut), to finally the types with vessels surrounded by parenchyma (Fig. 3.11). Vessels within dead tissue such as tracheids can be regarded as most vulnerable, vessels surrounded (and thereby protected) by living parenchyma, are less vulnerable to fungal attack. Superimposed upon this is the much greater vulnerability of ring-porous trees in which the water-conducting vessels are very superficially located in the tree stem and because vessels are very large, efficient and therefore less numerous (Chap. 1.4). It has been mentioned before that elms with smaller vessels have been found to be less vulnerable to the Dutch elm disease than those with larger vessels (Elgersma 1970).

Vapor blockage of vessels has received very little attention from investigators studying vascular wilt diseases. It is interesting to read the literature in view of this, and look for circumstatial evidence, but there are usually some critical pieces of information missing. Exceptions are rather rare. Thus, Mathre (1964) described the pathogenicity of *Ceratocystis* species to *Pinus ponderosa*. He found that water conduction failed in an area beyond the visible infection site. This is most easily explained by vapor-blocked tracheids. Coutts (1976) reported dry zones in sapwood that resulted from fungal infections.

Although I have made a very strong case for vapor blockage on these pages, this does not mean that I do not believe other types of blockage are possible under certain circumstances, especially when xylem pressures are high. Vapor blockage is undoubtedly the first and thereby real cause of failure of water transport in many cases, especially in ring-porous trees. But it is, of course, possible that certain pathogens, especially fungi, manage to enter vessels that contain water under tension. In fact, some evidence of this has been obtained by Newbanks (personal communication) when he found hyphae of *Verticillium* in vessels of *Acer saccharum* that had conducted dye, taken up from a vacuum container. But as *Acer* has a very closely knit vessel network, contact-staining has not yet been entirely ruled out. How long tensile water can exist after hyphal penetration is unknown. We must assume that the hyphae keep consuming wall material and that the wall sooner or later becomes leaky.

An interesting case is the *Phialophora (Verticillium) cinerescens* disease of carnation (*Dianthus* sp.). The fungus enters the roots without visible injury and causes simultaneous yellowing of leaves and gummosis of vessels (Péresse and Moreau 1968). The water status of the leaves at the time of yellowing seems to be unknown. If, for example yellowing were a senescence effect, caused by the closure of stomata, triggered by a xylem solute from the infection site (as described above for lethal yellowing of coconut palms), pressures would remain high in the xylem, the fungus could enter vessels without air seeding. We would then have a case where vessels are damaged beyond repair, but the damage would remain undetectable, because of pathologically induced high xylem pressure. Teleologically speaking, if the fungus induces the stomata to close first, it can send its spores up (but not down) in the intact, residual transpiration stream! Likewise, plants grown under very favorable conditions in the greenhouse, or in rainy weather outdoors, may possibly not show xylem damage due to unusually high xylem pressures. Such situations can be quantitatively considered by monitoring xylem pressure and applying Eq. (3.1).

In summary, it is perfectly conceivable that in a plant whose xylem undergoes the normal diurnal pressure fluctuation, i.e., experiences low pressures during the day, the xylem is easily made "leaky" by pathogen action, hence vessels embolize easily and are sealed off by gums and tyloses. If, on the other hand, xylem pressure is maintained high by external factors such as persistent rain, identical action of the pathogen would not embolize the vessels, the protective tylosis formation and or gummosis would not take place in water-filled vessels and might provide greater distribution of the pathogen. By the same token, embolism is somewhat less likely in root than in shoot infections, because pressures are always higher in the roots. While such considerations are purely theoretical, they show infection processes from new angles, and may suggest new approaches for the investigation of xylem dysfunction.

7.5 Pathogen Effects on Xylem Differentiation

In Chap. 6.3 we have already seen examples of the plant's possible reaction to the presence of pathogens in the xylem: the secretion of gums and the growth of tyloses from neighboring living parenchyma cells. As discussed in Chap. 6.3, all evidence available so far indicates that these events *follow* embolism. At least I am not aware of any firm evidence to the contrary. This is evidently a mechanism by which the plant limits the spread of pathogens. It is interesting that some authors suggest that gums are the result of vessel wall lysis; other authors, on the other hand, consider them a product of secretion from neighboring parenchyma cells (Catesson et al. 1976; Moreau et al. 1978). Whether gummosis can precede embolism in this case is not known.

Infection of carnations (*Dianthus* sp.) by *Phialophora cinerescens* also influences the differentiation of young xylem elements. While in healthy control plants membranes of intervessel pits are rendered more permeable by the hydrolysis of certain polysaccharides during the course of differentiation, this hydrolysis is stopped under the influence of fungal attack. When these tracheary elements mature, vessel-to-vessel conductance is reduced (Catesson et al. 1972).

The biochemistry of cell-wall carbohydrates and its modification by pathogens has been summarized recently by Fincher and Stone (1982).

Damaged xylem is sealed off from the functioning parts in other ways, namely by a layer of what early German botanists called Schutzholz (protective wood). The apoplast of this barrier zone is made impermeable by impregnation of the cell walls. Mullick (1977) found that the bordered pits of coniferous tracheids can become encrusted and thus non-conducting 19 days after the nearby bark has been wounded. This is interesting because it appears to be independent of embolism. Another reaction to xylem wounding, also non-specific, is the production of living parenchyma cells by the cambium which can thus effectively isolate new from old, injured or necrotic layers of xylem (Tippet and Shigo 1981).

7.6 The Problem of Injecting Liquids

Injecting chemicals such as fungicides into trees is a purely practical problem that actually does not belong into this book. The reason why a little section is devoted to injection is to summarize briefly the relevant anatomical and physiological properties that have been discussed before and are absolutely essential if one wants to make a successful injection.

First of all, it is worthwhile to study the dye ascent patterns in the tree species one is concerned with (Chap. 2.1). Each species has its characteristic tangential spread which usually amounts to an angle of 1°–3°. Obviously, the injected solution will spread more around the trunk, the longer the axial distance available along the trunk. For this reason, fewer injection holes are needed around the trunk, the taller the trunk and the lower the injection holes are placed on it. On the one hand we probably want to supply the entire crown with liquid, on the other we want to drill as few injection holes as possible, because each one injures the stem.

The question where to inject does not depend upon desired (maximum) distribution vs. (minimum) injury alone, it also depends upon where the tree is located. In an orchard tree we can do as we please, along a city street we have to consider the presence of pedestrians. For various reasons, small roots are the best injection points. Roots are relatively "disposable" organs of a tree, injuries to roots are in many cases less serious than injuries to the stem. Root diameter can be chosen so that rubber tubing can be fitted over them. The entire transverse section is thus available to injection, which is particularly important in ring-porous species. This brings us to the second important point.

We should know the extent of water conduction in the sapwood of the species we are dealing with. Diffuse-porous species such as most fruit trees, are no problem, because many growth rings of the sapwood contain conducting vessels and each injection point can thus reach a portion of the crown. The real difficulty arises with ring-porous species such as elms (*Ulmus* sp.), where it is almost exclusively the earlywood vessels of the most recent growth ring that conduct. Root injection is then particularly useful. Trunk injection is difficult to accomplish successfully, because the injection nozzle must be shaped in such a way (e.g., with lateral slits) that the liquid reaches the most recent growth ring which is located immediately beneath the cambium. It obviously makes no sense to drive an ordinary nozzle into the trunk of elm, thus blocking the conducting vessels, and then pumping gallons of expensive liquid into the central, non-conducting part of the wood. Radial xylem paths are minimally developed.

The third point to remember is the fact that xylem water is usually under tension during the day. When the hole is drilled, and the drill bit removed from the wood, air is being pulled into the ruptured vessels, thus blocking the xylem path. Subsequent injection may suffer from the air pockets between the applied liquid and the intact vessels beyond the injured ones. This problem has been discussed in Chap. 7.4. Species differ considerably in the efficiency of the extravascular water path from the applied liquid to the intact vessels. Vacuum infiltration immediately following drilling solves this problem partly, but I am not aware that anyone has used this method for therapeutic purposes, although we often use this method for experimental purposes. The advantage of vacuum infiltration is that there are small, hand-operated vacuum pumps available (no electricity needed). An alternative way of decreasing the bubble size is to apply the liquid under positive pressure. The air block diminishes in size when positive pressure is applied, according to the gas equation (the bubble size is inversely proportional to the absolute pressure). Some of these problems have been investigated systematically by Stephen Day for a Master's Thesis at the University of Maine in Orono (Day 1980).

The fourth problem is the pH of the injected solution. This has also been discussed briefly before. Alkaline solutes are absorbed by the vessel walls and have, as a result, very poor mobility. In other words, the solvent water moves much faster in the xylem than the solute. This is an advantage if one wants to mark the path of water, because subsequent dissection does not cause a loss of the dye, the dye track is permanently marked and can be observed after sectioning. An acid solute, on the other hand, is not absorbed on the vessel walls, i.e., it moves much faster and is better distributed. A pH slightly below neutral would probably be ideal. Unfortunately, some of the fungicides used against Dutch elm disease have such a low pH that they damage the xylem (see Newbanks et al. 1982).

Finally, especially when dealing with trees with very long vessels, such as elms, we must realize that injected liquid may never reach the visibly diseased parts of the crown, because much of the xylem path leading to the diseased parts of the crown may already be non-functional. Similarly, we must realize that when, during sanitation work, dead branches are removed, it may be necessary to remove a good portion of the proximal healthy-looking part of the branch as well. Cavitation carries fungal spores downward for considerable distances in species with long vessels such as elm. It is therefore very likely that fungal spores are already at the basal end of embolized vessels long before any symptom can be seen at that point. For practical purposes it is best to follow the advice of cooperative extension agents or experienced arborists who have learned by trial and error how much of the living part should be removed.

The safest test for the efficiency of an injection method is the chemical analysis of actual crown parts. Such work has been done by D.N. Roy and collaborators in Canada (see Newbanks et al. 1982).

It is obviously neither possible to give brief instructions to make therapeutic injections successful, nor is it the purpose of this book to give advice for such practical matters. In medicine, it is assumed that the operating surgeon is fully trained in human anatomy, physiology, pathology, etc. It would be far too expensive to have fully trained tree physiologists treat trees in backyards and on roadsides. But it is hoped that the few points made in this last section will help to point out the importance of understanding the tree's structure and function when one deals with practical problems.

Epilogue

This book has a rather strange and certainly a very brief history. During January, 1982, Tore Timell wrote me a letter asking if I would write a volume on dicotyledonous wood anatomy for his new series. My answer was no. First, I am not really that interested in wood anatomy for its own sake; second, I thought of *Trees, Structure and Function*, which should be revised. Tore answered immediately giving me free choice of title and time limit. This set me thinking and I realized that what I really wanted to write about was *Xylem Structure and the Ascent of Sap* (this is the final title; I started with the choice of about ten!)

I began on a trial basis about February 1, but became so fascinated that I did not stop until nine months later when the seven chapters were finished. Meanwhile, I had many colleagues read parts or all of the text. The manuscript was sent to the publisher on October 19 (and minor revisions continued until sometime in January). The manuscript was clean enough that galley proofs could be handled internally by the publisher.

I corrected page proofs in February and March. During that time I had trouble with my memory but I ascribed this to a rather nasty kidney stone affair I had experienced in January. One week after the last page proofs had gone back to the printer, the first CAT scan showed that, in fact, I probably have a tumor in the brain. At the time of this writing I do not know how long I'll have yet to live, but I do know that it may not be very long. The timing and incredibly fast writing of this book (about fourteen months from the inception of the idea to the end of printing) looks like a miracle to me.

MARTIN H. ZIMMERMANN

March 27, 1983

References

Addicott FT 1978 Abscission strategies in the behavior of tropical trees. In: Tomlinson PB, Zimmermann MH (eds) Tropical trees as living systems. Cambridge Univ Press, Cambridge London New York Melbourne, pp 381–398

Andrews HN 1961 Studies in paleobotany. John Wiley, New York, 487 pp

Apfel RE 1972 The tensile strength of liquids. Sci Am 227(2): 58–71

Arber A 1920 Water plants – a study of aquatic angiosperms. Cambridge Univ Press, Cambridge. Reprinted by Cramer, Weinheim, Germany, Wheldon & Wesley, Hafner Codicote Herts, New York, 436 pp

Ayres PG 1978 Water relations of diseased plants. In: Kozlowski TT (ed) Water deficits and plant growth Vol 5. Academic Press, New York London, pp 1–60

Baas P 1976 Some functional and adaptive aspects of vessel member morphology. Leiden Bot Ser 3: 157–181

Baas P (ed) 1982a New perspectives in wood anatomy. Nijhoff, Junk, The Hague, 252 pp

Baas P 1982b Systematic, phylogenetic, and ecological wood anatomy. In: Baas P (ed) New perspectives in wood anatomy. Nijhoff, Junk, The Hague, pp 23–58

Baas P, Zweypfenning RCVJ 1979 Wood anatomy of the Lythraceae. Acta Bot Neerl 28: 117–155

Bailey IW 1913 The preservative treatment of wood II. For Q 11: 12–20 (Reprinted as Chap 19 in: Bailey IW, Contributions to Plant Anatomy. Chron Bot Waltham Massachusetts 1954)

Bailey IW 1933 The cambium and its derivative tissues. VIII. Structure, distribution, and diagnostic significance of vestured pits in dicotyledons. J Arnold Arbor 14: 259–273

Bailey IW 1953 Evolution of the tracheary tissue of land plants. Am J Bot 40: 4–8 (Reprinted as Chap 13 in: Bailey IW (ed) Contributions to Plant Anatomy. Chron Bot Waltham Massachusetts 1954)

Bailey IW 1958 The structure of tracheids in relation to the movement of liquids, suspensions, and undissolved gases. In: Thimann KV (ed) The physiology of forest trees. Ronald, New York, pp 71–82

Bailey IW, Tupper WW 1918 Size variation in tracheary cells: I. A comparison between the secondary xylems of vascular cryptogams, gymnosperms, and angiosperms. Am Acad Arts Sci Proc 54: 149–204

Banks HP 1964 Evolution and plants of the past. Wadsworth, Belmont, 170 pp

Barghoorn ES 1964 Evolution of cambium in geologic time. In: Zimmermann MH (ed) The formation of wood in forest trees. Academic Press, New York London, pp 3–17

Bauch J 1973 Biologische Eigenschaften des Tannennaskerns. Mitt Bundesforschungsanst Forst Holzwirtsch 93: 213–224

Bauch J, Schweers W, Berndt H 1974 Lignification during heartwood formation: comparative study of rays and bordered pit membranes in coniferous woods. Holzforsch 28: 86–91

Begg JE, Turner NC 1970 Water potential gradients in field tobacco. Plant Physiol 46: 343–346

Bell A 1980 The vascular pattern of a rhizomatous ginger (Alpinia speciosa L, Zingiberaceae). Ann Bot 46: 203–220

Berger W 1931 Das Wasserleitungssystem von krautigen Pflanzen, Zwergsträuchern und Lianen in quantitativer Betrachtung. Beih Bot Cbl 48(I): 363–390

Berthelot M 1850 Sur quelques phénomènes de dilatation forcée des liquides. Ann Chim Phys 3e Sér 30: 232–237

Bierhorst DW 1958 Vessels in Equisetum. Am J Bot 45: 534–537

Bierhorst DW, Zamora PM 1965 Primary xylem elements and element associations of angiosperms. Am J Bot 52: 657–710

Bode HR 1923 Beiträge zur Dynamik der Wasserbewegungen in den Gefäßpflanzen. Jahrb Wiss Bot 62: 91–127

Böhm J 1867 Über die Funktion und Genesis der Zellen in den Gefäßen des Holzes. Sitzber Akad Wiss Vienna II Abt 55: 851

Böhm J 1893 Capillarität und Saftsteigen. Ber Dtsch Bot Ges 11: 203–212

Bolton AJ, Jardine P, Jones GL 1975 Interstitial spaces. A review and observations on some Araucariaceae. IAWA Bull 1975/1: 3–12

Bormann FH, Berlyn G (eds) 1981 Age and growth rate of tropical trees: new directions for research. Yale Univ School For Environ Stud Bull No 94, 137 pp

Bosshard HH 1955 Zur Physiologie des Eschenbraunkerns. Schweiz Z Forstwes 106: 592–612

Bosshard HH 1974 Holzkunde. Vol 1 und 2. Birkhäuser, Basel Stuttgart, 224 pp, 312 pp

Bosshard HH 1976 Jahrringe and Jahrringbrücken. Schweiz Z Forstwes 127: 675–693

Bosshard HH, Kučera L 1973 Die dreidimensionale Strukturanalyse des Holzes. I. Die Vernetzung des Gefäßsystems in Fagus sylvatica L. Holz Roh Werkst 31: 437–445

Bosshard HH, Kučera L, Stocker U 1978 Gewebe-Verknüpfungen in Quercus robur L. Schweiz Z Forstwes 129: 219–242

Botosso PC, Gomes AV 1982 Radial vessels and series of perforated ray cells in Annonaceae. IAWA Bull 3: 39–44

Boyer JS 1967 Leaf water potentials measured with a pressure chamber. Plant Phys 42: 133–137

Braun HJ 1959 Die Vernetzung der Gefäße bei Populus. Z Bot 47: 421–434

Braun HJ 1963 Die Organisation des Stammes von Bäumen und Sträuchern. Wissenschaft Verl Ges, Stuttgart, 162 pp

Braun HJ 1970 Funktionelle Histologie der sekundären Sproßachse. I Das Holz. In: Zimmermann W, Ozenda P, Wulff HD (eds) Encycl Plant Anat, 2 nd edn. Bornträger, Berlin Stuttgart, 190 pp

Breitsprecher A, Hughes W 1975 A recording dendrometer for humid environments. Biotropica 7: 90–99

Briggs LJ 1950 Limiting negative pressure of water. J Appl Phys 21: 721–722

Bristow JM 1975 The structure and function of roots in aquatic vascular plants. In: Torrey JG, Clarkson DT (eds) The development and function of roots. Academic Press, New York London, pp 221–236

Buchholz M 1921 Über die Wasserleitungsbahnen in den interkalaren Wachstumszonen monokotyler Sprosse. Flora 114: 119–186

Budgett HM 1912 The adherence of flat surfaces. Proc R Soc Lond A 86: 25–35

Burger H 1953 Holz, Blattmenge und Zuwachs. XIII. Fichten im gleichaltrigen Hochwald. Mitt Schweiz Anst Forst Versuchswes 9: 38–130

Burggraaf PD 1972 Some observations on the course of the vessels in the wood of Fraxinus excelsior L. Acta Bot Neerl 21: 32–47

Burke MJ, Gusta LV, Quamme HA, Weiser CJ, Li PH 1976 Freezing and injury in plants. Annu Rev Plant Physiol 27: 507–528

Butterfield BG, Meylan BA 1982 Cell wall hydrolysis in the tracheary elements of the secondary xylem. In: Baas P (ed) New perspectives in wood anatomy. Nijhoff, Junk, The Hague, pp 71–84

Campana RJ 1978 Inoculation and fungal invasion of the tree. In: Sinclair WA, Campana RJ (eds) Dutch elm disease: perspectives after 60 years. Cornell Univ Agric Expt Stn Search Agric 8(5): 17–20

Carlquist S 1975 Ecological strategies of xylem evolution. Univ Calif Press, Berkeley Los Angeles London, 259 pp

Carte AE 1961 Air bubbles in ice. Proc Physi Soc 77: 757–768

Carter JC 1945 Wetwood of elms. Nat Hist Survey 23: 407–448

Catesson AM, Czaninski Y, Moreau M, Péresse M 1979 Conséquences d'une infection vasculaire sur la maturation des vaisseaux. Rev Mycol 43: 239–243

Catesson AM, Czaninski Y, Péresse M, Moreau M 1972 Modifications des parois vasculaires de l'oeillet infecté par le Phialophora cinerescens (Wr) van Beyma. CR Acad Sci Paris Ser D 275: 827–829

Catesson AM, Czaninski Y, Péresse M, Moreau M 1976 Sécrétions intravasculaires de substances „gommeuses" par les cellules associées aux vaisseaux en réaction à une attaque parasitaire. Soc Bot Fr Coll Sécrét Végét 123: 93–107

Chattaway MM 1948 Note on the vascular tissue in the rays of Banksia. J Counc Sci Ind Res 21: 275–278

Chattaway MM 1949 The development of tyloses and secretion of gum in heartwood formation. Aust J Sci Res Ser B Biol Sci 2: 227–240

Chattaway MM 1952 The sapwood-heartwod transition. Aust For 16: 25–34

Cheadle VI 1953 Independent origin of vessels in the monocotyledons and dicotyledons. Phytomorphology 3: 23–44

Chen PYS, Sucoff EI, Hossfeld R 1970 The effect of cations on the permeability of wood to aqueous solutions. Holzforsch 24: 65–67

Chou CK, Levi MP 1971 An electron microscopical study of the penetration and decomposition of tracheid walls of Pinus sylvestris by Poria vaillantii. Holzforsch 25: 107–112

Cichan MA, Taylor TN, 1982 Vascular combium development in Sphenophyllum: a carboniferous arthrophyte. LWA Bull 3: 155–160

Clarkson DT, Robards AW 1975 The endodermis, its structural development and physiological role. In: Torrey JG, Clarkson DT (eds) The development and function of roots. Academic Press, New York London, pp 415–436

Cohen Y, Fuchs M, Green GC 1981 Improvement of the heat pulse method for determining sap flow in trees. Plant Cell Environ 4: 391–397

Comstock GL, Côté WA 1968 Factors affecting permeability and pit aspiration in coniferous sapwood. Wood Sci Technol 2: 279–291

Connor DJ, Legge NJ, Turner NC 1977 Water relations of mountain ash (Eucalyptus regnans F Muell) forests. Aust J Plant Physiol 4: 735–762

Conway VM 1940 Growth rates and water loss in Cladium mariscus R Br. Ann Bot NS 4: 151–164

Core HA, Côté WA Jr, Day AC 1979 Wood structure and identification, 2nd edn. Syracuse Univ Press, Syracuse, 172 pp

Coster C 1927 Zur Anatomie und Physiologie der Zuwachszonen- und Jahresringbildung in den Tropen. Ann Jardin Bot Buitenzorg 38: 1–114

Côté WA Jr, Day AC 1962 Vestured pits – fine structure and apparent relationship with warts. Tappi 45: 906–910

Coutts MP 1976 The formation of dry zones in the sapwood of conifers I. Europ J For Pathol 6: 372–381

Coutts MP, Rishbeth J 1977 The formation of wetwood in grand fir. Eur J For Pathol 7: 13–22

Day SJ 1980 The influence of sapstream continuity and pressure on distribution of systemic chemicals in American elm (Ulmus americana L). Master's Thesis, Univ Maine, Orono

Dimond AE 1966 Pressure and flow relations in vascular bundles of the tomato plant. Plant Physiol 41: 119–131

Dimond AE 1970 Biophysics and biochemistry of the vascular wilt syndrome. Annu Rev Phytopathol 8: 301–322

Dixon HH 1914 Transpiration and the ascent of sap in plants. MacMillan, London, 216 pp

Dobbs RC, Scott DRM 1971 Distribution of diurnal fluctuations in stem circumference of Douglas fir. Can J For Res 1: 80–83

Donny J 1846 Sur la cohésion des liquides et sur leur adhésion aux corps solides. Ann Chim Phys 3e Sér 16: 167

Duchartre P 1858 Recherches expérimentales sur la transpiration des plantes dans les milieux humides. Bull Soc Bot France 5: 105–111

Elgersma DW 1970 Length and diameter of xylem vessels as factors in resistance of elms to Ceratocystis ulmi. Neth J Plant Path 76: 179–182

Esau K 1965 Plant anatomy, 2nd edn. John Wiley, New York, 767 pp

Eschrich W 1975 Sealing systems in phloem. In: Zimmermann MH, Milburn JA (eds) Transport in Plants I. Encyclopedia of Plant Physiology New Ser Vol 1. Springer, Berlin Heidelberg New York, pp 39–56

Ewart AJ 1905–1906/1907–1908 The ascent of water in trees. Philos Trans Soc Lond B 198: 41–45, 199: 341–392

Faber von FC 1915 Physiologische Fragmente aus einem tropischen Urwald. Jb Wiss Bot 56: 197–220

Fahn A 1964 Some anatomical adaptations of desert plants. Phytomorphology 14: 93–102

Fahn A 1974 Plant anatomy, 2nd edn. Pergamon, Oxford, 611 pp

Farmer JB 1918 On the quantitative differences in the water conductivity of the wood in trees and shrubs. Proc Soc Lond B 90: 218–250

Fegel AC 1941 Comparative anatomy and varying physical properties of trunk, branch, and root wood in certain northeastern trees. Bull NY State Coll For Syracuse Univ Vol 14 No 2b Tech Publ No 55: 1–20

Fincher GB, Stone BA 1981 Metabolism of non-cellulosic polysaccharides. In: Tanner W, Loewus FA (eds) Plant carbohydrates II. Encyclopedia of Plant Physiology New Ser Vol 13 B. Springer, Berlin Heidelberg New York, pp 68–132

Filzner P 1948 Ein Beitrag zur ökologischen Anatomie von Rhynia. Biol Zbl 67: 13–17

Firbas F 1931a Untersuchungen über den Wasserhaushalt der Hochmoorpflanzen. Jb Wiss Bot 74: 459–696

Firbas F 1931b Über die Ausbildung des Leitungssystems und das Verhalten der Spaltöffnungen im Frühjahr bei Pflanzen des Mediterrangebietes und der tunesischen Steppen und Wüsten. Beih Bot Cbl 48: 451–465

Foster AS 1956 Plant idioblasts: remarkable examples of cell specialization. Protoplasma 46: 183–193

French JC, Tomlinson PB 1981a Vascular patterns in stems of Araceae: subfamilies Calloideae and Lasioideae. Bot Gaz 142: 366–381

French JC, Tomlinson PB 1981b Vascular patterns in stems of Araceae: subfamily Pothoideae. Am J Bot 68: 713–729

French JC, Tomlinson PB 1981c Vascular patterns in stems of Araceae: subfamily Monsteroideae. Am J Bot 68: 1115–1129

French JC, Tomlinson PB 1981d Vascular patterns in stems of Araceae: subfamily Philodendroideae. Bot Gaz 142: 550–563

Frenzel P 1929 Über die Porengrößen einiger pflanzlicher Zellmembranen. Planta 8: 642–665

Frey-Wyssling A 1959 Die pflanzliche Zellwand. Springer, Berlin Göttingen Heidelberg, 367 pp

Frey-Wyssling A 1982 Introduction to the symposium: cell wall structure and biogenesis. IAWA Bull 3: 25–30

Frey-Wyssling A, Bosshard HH 1953 Über den Feinbau der Schließhäute in Hoftüpfeln. Holz Roh Werkst 11: 417–420

Frey-Wyssling A, Mühlethaler K, Bosshard HH 1955 Das Elektronenmikroskop im Dienste der Bestimmung von Pinusarten. Holz Roh Werkst 13: 245–249, 14: 161–162 (1956)

Frey-Wyssling A, Mühlethaler K, Bosshard HH 1959 Über die mikroskopische Auflösung der Haltefäden des Torus in Hoftüpfeln. Holzforsch Holzverwert 11: 107–108

Friedrich J 1897 Über den Einfluß der Witterung auf den Baumzuwachs. Zbl ges Forstwes 23: 471–495

Fritts HC, Fritts EC 1955 A new dendrograph for recording radial changes of a tree. For Sci 1: 271–276

Fujita M, Nakagawa K, Mori N, Harada H 1978 The season of tylosis development and changes in parenchyma cell structure in Robinia pseudoacacia L. Bull Kyoto Univ For 50: 183–190

Gardiner W 1883 On the physiological significance of water glands and nectaries. Proc Cambridge Philos Soc 5: 35–50

Gessner F 1951 Untersuchungen über den Wasserhaushalt der Nymphaeaceen. Biol Generalis 19: 247–280

Gibbs RD 1958 Patterns of the seasonal water content of trees. In: Thimann KV (ed) The physiology of forest trees. Ronald, New York, pp 43–69

van der Graaff NA, Baas P 1974 Wood anatomical variation in relation to latitude and altitude. Blumea 22: 101–121

Greenidge KNH 1952 An approach to the study of vessel length in hardwood species. Am J Bot 39: 570–574

Greenidge KNH 1957 Ascent of sap. Annu Rev Plant Physiol 8: 237–256

Greenidge KNH 1958 Rates and patterns of moisture movement in trees. In: Thimann KV (ed) The physiology of forest trees. Ronald, New York, pp 19–41

Grier CC, Waring RH 1974 Conifer foliage mass related to sapwood area. For Sci 20: 205–206

Haberlandt G 1914 Physiological plant anatomy. Translated from the 4th German edn by Drummond M, MacMillan, London, 777 pp

Haider K 1954 Zur Morphologie und Physiologie der Sporangien leptosporangiater Farne. Planta 44: 370–411

Hallé F, Oldeman RAA, Tomlinson PB 1978 Tropical trees and forests, an architectural analysis. Springer, Berlin Heidelberg New York, 420 pp

Hammel HT 1967 Freezing of xylem sap without cavitation. Plant Physiol 42: 55–66

Hammel HT 1968 Measurement of turgor pressure and its gradient in the phloem of oak. Plant Physiol 43: 1042–1048

Hammel HT, Scholander PF 1976 Osmosis and tensile solvent. Springer, Berlin Heidelberg New York, 133 pp

Handley WRC 1936 Some observations on the problem of vessel length determination in woody dicotyledons. New Phytol 35: 456–471

Harlow WM 1970 Inside wood, masterpiece of nature. Am For Assoc, Washington DC, 120 pp

Hartig R 1889 Ein Ringelungsversuch. Allg Forst Jagdztg pp 365–373, pp 401–410

Hartig Th 1878 Anatomie und Physiologie der Holzpflanzen. Springer, Berlin, 412 pp

Hartley C, Davidson RW, Crandall BS 1961 Wetwood, bacteria, and increased pH in trees. US For Prod Lab Report 2215: 1–34

Hatheway WH, Winter DF 1981 Water transport and storage in Douglas fir: a mathematical model. Mitt Forst Bundes Versuchsanst Wien 142: 193–222

Häusermann E 1944 Über die Benetzungsgröße der Mesophyllinterzellularen. Ber Schweiz Bot Ges 54: 541–578

Hejnowicz Z, Romberger JA 1973 Migrating cambial domains and the origin of wavy grain in xylem of broadleaved trees. Am J Bot 209–222

Hellqvist J, Richards GP, Jarvis PG 1974 Vertical gradients of water potential and tissue water relations in Sitka spruce trees measured with the pressure chamber. J Appl Ecol 11: 637–668

Hillis WE 1977 Secondary changes in wood. In: Loewus FA, Runeckles VC (eds) The structure, biosynthesis, and degradation of wood. Plenum, New York London, pp 247–309

Höll W 1975 Radial transport in rays. In: Zimmermann MH, Milburn JA (eds) Transport in plants I. Encyclopedia of Plant Physiology New Ser Vol 1. Springer, Berlin Heidelberg New York, pp 432–450

Holle H 1915 Untersuchungen über Welken, Vertrocknen und Wiederstraffwerden. Flora 108: 73–126

Hong SG, Sucoff E 1982 Rapid increase in deep supercooling of xylem parenchyma. Plant Physiol 69: 697–700

Hook DD, Brown CL, Wetmore RH 1972 Aeration in trees. Bot Gaz 133: 443–454

Howard FW, Thomas DL 1980 Transmission of palm lethal decline to Veitchia merrillii by a planthopper Myndus crudus. J Encon Entomol 73: 715–717

Howard FW, Thomas DL, Donselman HM, Collins ME 1979 Susceptibilities of palm species to mycoplasmalike organism-associated diseases in Florida. Plant Pro Bull 27: 109–117

Howard RA 1974 The stem-node-leaf continuum of the Dicotyledonae. J Arnold Arbor 55: 125–181

Huber B 1928 Weitere quantitative Untersuchungen über das Wasserleitungssystem der Pflanzen. Jb Wiss Bot 67: 877–959

Huber B 1932 Beobachtung und Messung pflanzlicher Saftströme. Ber Dtsch Bot Ges 50: 89–109

Huber B 1935 Die physiologische Bedeutung der Ring- und Zerstreutporigkeit. (Physiological significance of ring- and diffuse-porousness). Ber Dtsch Bot Ges 53: 711–719 (Xerox copy of English translation available from National Translation Center, 35 West 33 rd St, Chicago, IL 60616 USA)

Huber B 1956 Die Gefäßleitung. In: Ruhland W (ed) Encyclopedia of Plant Physiology Vol 3. Springer, Berlin Göttingen Heidelberg, pp 541–582

Huber B, Merz W 1958 Über die Bedeutung des Hoftüpfelverschlusses für die axiale Wasserleitfähigkeit von Nadelhölzern. Planta 51: 645–672

Huber B, Schmidt E 1936 Weitere thermo-elektrische Untersuchungen über den Transpirationsstrom der Bäume. (Further thermo-electric investigations on the transpiration stream in trees) Tharandt Forst Jb 87: 369–412 (Xerox copies of English translation available from National Translation Center, 35 West 33 rd St, Chicago, IL 60616 USA)

Huber B, Schmidt E 1937 Eine Kompensationsmethode zur thermo-elektrischen Messung langsamer Saftströme. Ber Deutsch Bot Ges 55: 514–529 (Xerox copies of English translation available from National Translater Center, 35 West 33 rd St, Chicago, IL 60616 USA)

Hudson MS, Shelton SV 1969 Longitudinal flow of liquids in southern pine poles. For Prod J 19: 25–32

Ipsen JD, Abul-Haji YJ 1982 Fluorescent antibody technique as a means of localizing Ceratocystis ulmi toxins in elm. Can J Bot 60: 724–729

Isebrands JG, Larson PR 1977 Vascular anatomy of the nodal region in eastern cottonwood. Am J Bot 64: 1066–1077

Jaccard P 1913 Eine neue Auffassung über die Ursachen des Dickenwachstums. Naturwiss Z Forst Landwirtsch 11: 241–279

Jaccard P 1919 Nouvelles recherches sur l'accroissement en épaisseur des arbres. Publ No 23 Fondation Schnyder von Wartensee, Zürich

Jeník J 1978 Discussion. In: Tomlinson PB, Zimmermann MH (eds) Tropical trees in living systems. Cambridge Univ Press, Cambridge London New York Melbourne, 529 p

Jeje AA, Zimmermann MH 1979 Resistance to water flow in xylem vessels. J Exp Bot 30: 817–827

Jones CS, Lord EM 1982 The development of split axes in Ambrosia dumonsa (Gray) Payne (Asteraceae). Bot Gaz 143: 446–453

Jost L 1916 Versuche über die Wasserleitung in der Pflanze. Z Bot 8: 1–55

Kaiser P 1879 Über die tägliche Periodizität der Dickendimensionen der Baumstämme. Inaug Diss, Halle, 38 pp

Kaufman MR 1968 Evaluation of the pressure chamber technique for estimating plant water potential of forest tree species. For Sci 14: 369–374

Kelso WC, Gertjejansen CO, Hossfeld RL 1963 The effect of air blockage upon the permeability of wood to liquids. Univ Minnesota Agric Res Stn Tech Bull 242: 1–40

Klein G 1923 Zur Aetiologie der Thyllen. Z Bot 15: 418–439

Klepper B, Browning VD, Taylor HM 1971 Stem diameter in relation to plant water status. Plant Physiol 48: 683–685

Klotz LH 1978 Form of the perforation plates in the wide vessels of metaxylem in palms. J Arnold Arbor 59: 105–128

Kórán Z, Côté WA 1965 The ultrastructure of tyloses. In: Côté WA (ed) Cellular ultrastructure of woody plants. Syracuse Univ Press, Syracuse NY, pp 319–333

Krahmer RL, Côté WA 1963 Changes in coniferous wood cells associated with heartwood formation. Tappi 46: 42–49

Kramer K 1974 Die tertiären Hölzer Südost-Asiens, I und II. Palaeontogr Abt B 144: 45–181, 145: 1–150

Kramer PJ 1937 The relation between the rate of transpiration and the rate of absorption of water in plants. Am J Bot 24: 10–15

Kraus G 1877 Die Verteilung und Bedeutung des Wassers bei Wachstums- und Spannungsvorgängen in der Pflanze. Bot Ztg 35: 595–597

Kraus G 1881 Die tägliche Schwellungsperiode der Pflanzen. Abhandl Naturf Gesellsch Halle 15

Kučera L 1975 Die dreidimensionale Strukturanalyse des Holzes. II. Das Gefäßstrahlnetz bei der Buche (Fagus sylvatica L). Holz Roh Werkst 33: 276–282

Kursanov AL 1957 The root system as an organ of metabolism. UNESCO Intern Conf Radioisotopes Sci Res, UNESCO/NS/RIC/128, Pergamon, London, 12 pp

Laming PB, ter Welle BJH 1971 Anomalous tangential pitting in Picea abies Karst. (European spruce) IAWA Bull 1971/4: 3–10

Larson PR, Isebrands JG 1978 Functional significance of the nodal constricted zone in Populus deltoides. Can J Bot 56: 801–804

Läuchli A, Bieleski RL 1983 Inorganic plant nutrition. Encyclopedia of Plant Physiology New Ser Vol 15. Springer, Berlin Heidelberg New York

Legge NJ 1980 Aspects of transpiration in mountain ash Eucalyptus regnans F Muell. PhD Thesis, LaTrobe Univ, Melbourne

Liese W, Bauch J 1964 Über die Wegsamkeit der Hoftüpfel von Coniferen. Naturwissenschaften 21: 516

Liese W, Ledbetter MC 1963 Occurrence of a warty layer in vascular cells of plants. Nature 197: 201–202

Lybeck BR 1959 Winter freezing in relation to the rise of sap in tall trees. Plant Physiol 34: 482–486

Mackay JFG, Weatherley PE 1973 The effects of transverse cuts through the stems of transpiring woody plants on water transport and stress in leaves. J Exp Bot 24: 15–28

MacDougal DT 1924 Dendrographic measurements. Carnegie Inst Washington Publ 350: 1–88

MacDougal DT, Overton JB, Smith GM 1929 The hydrostatic-pneumatic system of certain trees: movements of liquids and gases. Carnegie Inst Washington Publ 397: 1–99

Mantai KE, Newton ME 1982 Root growth in Myriophyllum: a specific plant response to nutrient availability? Aquatic Bot 13: 45–55

Marshall DC 1958 Measurement of sap flow in conifers by heat transport. Plant Physiol 33: 385–396

Martens P 1971 Un marronnier centenaire privé d'écorce à sa base depuis plus d'un demi-siècle. Bull Acad Roy Belgique Classe Sci 57: 65–84

Mathre DE 1964 Pathogenicity of Ceratocystis ips and Ceratocystis minor to Pinus ponderosa. Contrib Boyce Thompson Inst 22: 363–388

McCoy RE 1978 Mycoplasmas and yellows diseases. In: Barile MF, Razin S, Tully JG, Whitcomb RF (eds) The mycoplasmas Vol 2. Academic Press, New York London

McDonough J, Zimmermann MH 1979 Effect of lethal yellowing on xylem pressure in coconut palms. Principes 23: 132–137

Metzger K 1894, 1895 Studien über den Aufbau der Waldbäume und Bestände nach statischen Gesetzen. Mündener Forstl Hefte 5: 61–74 (1894), 6: 94–119 (1894), 7: 45–97 (1895)

Meyer FJ 1928 Die Begriffe „stammeigene Bündel" und „Blattspurbündel" im Lichte unserer heutigen Kenntnisse vom Aufbau und der physiologischen Wirkungsweise der Leitbündel. Jb Wiss Bot 69: 237–263

Meyer RW, Côté WA Jr 1968 Formation of the protective layer and its role in tylosis development. Wood Sci Technol 2: 84–94

Meylan BA, Butterfield BG 1972 Three-dimensional structure of wood. Chapman and Hall, London, 80 pp

Meylan BA, Butterfield BG 1978a The structure of New Zealand woods. NZ Dep Sci Ind Res Bull 222: 1–250

Meylan BA, Butterfield BG 1978b Occurrence of helical thickenings in the vessels of New Zealand woods. New Phytol 81: 139–146

Milburn JA 1973a Cavitation in Ricinus by acoustic detection: Induction in excised leaves by various factors. Planta 110: 253–265

Milburn JA 1973b Cavitation studies on whole Ricinus plants by acoustic detection. Planta 112: 333–342

Milburn JA, Johnson RPC 1966 The conduction of sap. II. Detection of vibrations produced by sap cavitation in Ricinus xylem. Planta 69: 43–52

Milburn JA, McLaughlin ME 1974 Studies of cavitation in isolated vascular bundles and whole leaves of Plantago major L. New Phytol 73: 861–871

Minden M von 1899 Beiträge zur anatomischen und physiologischen Kenntnis Wasser-sezernierender Organe. Bibl Bot 9(46): 1–76

Mittler TE 1967 Water tensions in plants – an entomological approach. Ann Entomol Soc Am 60: 1074–1076

Moreau M, Catesson AM, Péresse M, Czaninski Y 1978 Dynamique comparée des réactions cytologiques du xylème de l'Oeillet en présence de parasites vasculaires. Phytopath Z 91: 289–306

Mullick DB 1977 The non-specific nature of defense in bark and wood during wounding, insect and pathogen attack. In: Loewus FA, Runeckles VC (eds) The structure, biogenesis, and degradation of wood. Plenum, New York London, pp 395–442

Münch E 1930 Die Stoffbewegungen in der Pflanze. Fischer, Jena, 234 pp

Münch E 1943 Durchlässigkeit der Siebröhren für Druckströmungen. Flora 136: 223–262

Murdoch CW 1979 A selected bibliography on bacterial wetwood of trees. Univ Maine School For Res Tech Note 72: 1–7

Murdoch CW 1981 Bacterial wetwood in elm. PhD Thesis, Univ Maine, Orono (Plant Sci), 145 pp

Nakashima H 1924 Über den Einfluß meteorologischer Faktoren auf den Baumzuwachs. J Coll Agric (Tokyo) 12: 71–262

Newbanks D, Roy DN, Zimmermann MH 1982 Dutch elm disease: what an arborist should know. Arnoldia 42: 60–69

Newbanks D, Bosch A, Zimmermann MH 1983 Evidence for xylem dysfunction by embolization in Dutch elm disease. Phytopathology (in press)

O'Brien TP, Carr DJ 1970 A suberized layer in the cell walls of the bundle sheath of grasses. Aust J Biol Sci 23: 275–287

O'Brien TP, Thimann KV 1967 Observations on the fine structure of the oat coleoptile III. Protoplasma 63: 443–478

Oertli JJ 1971 The stability of water under tension in the xylem. Z Pflanzenphysiol 65: 195–209

Ohtani J, Ishida S 1976 Study on the pit of wood cells using scanning electron microscopy. Report 5. Vestured pits in Japanese dicotyledonous woods. Res Bull Coll Expt For Coll Agric Hokkaido Univ 33: 407–436

Olien WC, Bukovac MJ 1982 Ethephon-induced gummosis in sour cherry (Prunus cerasus L). Plant Physiol 70: 547–555, 556–559

Panshin AJ, de Zeeuw C 1980 Textbook of wood technology, 4th edn. McGraw-Hill, New York, 722 pp

Parthasarathy MV 1974 Mycoplasma-like organisms associated with lethal-yellowing disease of palms. Phytopathology 64: 667–674

Passioura JB 1972 The effect of root geometry on the yield of wheat growing on stored water. Aust J Agric Res 23: 745–752

Peck AJ, Rabbidge RM 1969 Design and performance of an osmotic tensiometer for measuring capillary potential. Soil Sci Soc Am Proc 33: 196–202

Péresse M, Moreau M 1968 Essai d'expression numérique des symptômes de la verticilliose sur oeillets. CR Soc Biol 162: 234

Petty JA 1978 Fluid flow through the vessels of birch wood. J Exp Bot 29: 1463–1469

Petty JA 1981 Fluid flow through the vessels and intervascular pits of sycamore wood. Holzforsch 35: 213–216

Pickard WF 1981 The ascent of sap in plants. Prog Biophys Molec Biol 37: 181–229

Pomerleau R, Mehran AR 1966 Distribution of spores of Ceratocystis ulmi labelled with phosphorus-32 in green shoots and leaves of Ulmus americana. Nat Gen Que 93: 577–582

Praël E 1888 Vergleichende Untersuchungen über Schutz- und Kernholz der Laubbäume. Jb Wiss Bot 19: 1–81

Preston RD 1952 Movement of water in higher plants. In: Frey-Wyssling (ed) Deformation and flow in biological systems. Elsevier North-Holland, Amsterdam, pp 257–321

Pridgeon AM 1982 Diagnostic anatomical characters in the Pleurothallidinae (Orchidaceae). Am J Bot 69: 921–938

Priestley JH, Scott LI, Malins ME, 1933 A new method of studying cambial activity. Proc Leeds Phil Soc 2: 365–374

Priestley JH, Scott LI, Malins ME, 1935 Vessel development in the angiosperms. Proc. Leeds Phil. Soc 3: 42–54

Raschke K 1975 Stomatal action. Annu Rev Plant Physiol 226: 309–340

Reichenbach H von 1845 Untersuchungen über die zellenartigen Ausfüllungen der Gefäße. (Published anonymously) Z Bot 3: 225–231, 241–253

Reicher K 1907 Die Kinematographie in der Neurologie. Neurol Zentralbl 26: 496

Rein H 1928 Die Thermo-Stromuhr. Ein Verfahren welches mit etwa ± 10 Prozent Genauigkeit die unblutige langdauernde Messung der mittleren Durchflußmengen an gleichzeitigen Gefäßen gestattet. Z Biol 87: 394–418

Reiner M 1960 Deformation, strain, and flow. An elementary introduction to rheology. Lewis, London, 347 pp

Renner O 1911 Experimentelle Beiträge zur Kenntnis der Wasserbewegung. Flora 103: 173–247

Renner O 1915 Theoretisches und Experimentelles zur Kohäsionetheorie der Wasserbewegung. Jb Wiss Bot 56: 617–667

Renner O 1925 Die Porenweite der Zellhäute und ihre Beziehung zum Saftsteigen. Ber Dtsch Bot Ges 43: 207–211

Richter H 1974 Erhöhte Saugspannungswerte und morphologische Veränderungen durch transversale Einschnitte in einen Taxus-Stamm. Flora 163: 291–309

Richter JP 1970 The notebooks of Leonardo da Vinci (1452–1519), compiled and edited from the original manuscripts. Dover, New York. (Reprint of a work originally published by Sampson Low Marston Searle and Rivington London 1883)

Riedl H 1937 Bau und Leistungen des Wurzelholzes. (Structure and function of root wood). Jb Wiss Bot 85: 1–75 (Xerox copy of English translation available from National Translation Center, 35 West 33 rd St, Chicago IL 60616 USA)

Rodger EA 1933 Wound healing in submerged plants. Am Midl Nat 14: 704–713

Rouschal E 1937 Die Geschwindigkeit des Transpirationsstromes in Macchiengehölzen (Thermoelektrische Messungen). Sitzgsber Akad Wiss Wien 146: 119–133

Rouschal E 1940 Fluoreszenzoptische Messungen der Geschwindigkeit des Transpirationsstromes an krautigen Pflanzen mit Berücksichtigung der Blattspurleitflächen. Flora 134: 229–256

Rouschal E 1941 Beiträge zum Wasserhaushalt von Gramineen und Cyperaceen I. Die fasziküläre Wasserleitung in den Blättern und ihre Beziehung zur Transpiration. Planta 32: 66–87

Rübel E 1919 Experimentelle Untersuchungen über die Beziehungen zwischen Wasserleitungsbahn und Transpirationsverhältnissen bei Helianthus annuus L. Beih Bot Cbl 37: 1–62

Rundel PW, Stecker RE 1977 Morphological adaptations of tracheid structure to water stress gradients in the crown of Sequoiadendron giganteum. Oecologia 27: 135–139

Salisbury EJ 1913 The determining factors in petiolar structure. New Phytol 12: 281–289

Sanio K 1872 Über die Größe der Holzzellen bei der gemeinen Kiefer (Pinus silvestris). Jb Wiss Bot 8: 401–420

Sauter JJ 1966 Untersuchungen zur Physiologie der Pappelholzstrahlen. II. Z Pflanzenphysiol 55: 349–362

Sauter JJ 1967 Der Einfluß verschiedener Temperaturen auf die Reservestärke in parenchymatischen Geweben von Baumsproßachsen. Z Pflanzenphysiol 56: 340–352

Sauter JJ 1976 Analysis of the amino acids and amides in the xylem sap of Salix caprea L in early spring. Z Pflanzenphysiol 79: 276–280

Sauter JJ 1980 Seasonal variation of sucrose content in the xylem sap of Salix. Z Pflanzenphysiol 98: 377–391

Sax K 1954 The control of tree growth by phloem blocks. J Arnold Arbor 35: 251–258

Schmid R, Machado RD 1968 Pit membranes in hardwoods – fine structure and development. Protoplasma 66: 185–204

Schneider IR, Worley JF 1959 Upward and downward transport of infectious particles of southern bean mosaic virus through steamed portions of bean stems. Virology 8: 230–242

Scholander PF 1958 The rise of sap in lianas. In: Thimann KV (ed) The physiology of forest trees. Ronald, New York, pp 3–17

Scholander PF 1972 Tensile water. Am Sci 60: 584–590

Scholander PF, Flagg W, Hock RJ, Irving L 1953 Studies on the physiology of frozen plants and animals in the arctic. J Cell Comp Physiol 42 Suppl 1: 1–56

Scholander PF, Love WE, Kanwisher JW 1955 The rise of sap in tall grapevines. Plant Physiol 30: 93–104

Scholander PF, Hemmingsen E, Garey W 1961 Cohesive lift of sap in the rattan vine. Science 134: 1835–1838

Scholander PF, Hammel HT, Bradstreet ED, Hemmingsen EA 1965 Sap pressures in vascular plants. Sci 148: 339–346

Schoute JC 1912 Über das Dickenwachstum der Palmen. Ann Jard Bot Buitenzorg 2e Ser 11: 1–209

Schwendener S 1890 Die Mestomscheiden der Gramineenblätter. Sitzungsber Preuss Akad Wiss Phys Math Kl 22: 405–426

Scott FM 1950 Internal suberization of tissues. Bot Gaz 111: 378

Sculthorpe CD 1967 The biology of aquatic vascular plants. Arnold, London, 610 pp

Shain L, Mackay JFG 1973 Seasonal fluctuation in respiration of aging xylem in relation to heartwood formation in Pinus radiata. Can J Bot 51: 737–741

Shibata N, Harada H, Saiki H 1981 Difference in the development of incubated tyloses within the sapwood of Castanea crenata Sieb. et Zucc. Bull Kyoto Univ For 53: 231–240

Shinozaki K, Yoda K, Hozumi K, Kira T 1964 A quantitative analysis of plant form – the pipe model theory I. Basic analyses II. Further evidence of the theory and its application in forest ecology. Jpn J Ecol 14: 97–105, 133–139

Sinclair WA, Campana RJ (eds) 1978 Dutch elm disease: perspectives after 60 years. Cornell Univ Agric Exp Stn Search Agric 8(5): 1–52

Skene DS, Balodis V 1968 A study of vessel length in Eucalyptus obliqua L'Hérit. J Exp Bot 19: 825–830

Snell K 1908 Untersuchungen über die Nahrungsaufnahme bei Wasserpflanzen. Flora 98: 213–249

Sovonick-Dunford S, Lee DR, Zimmermann MH 1981 Direct and indirect measurements of phloem turgor pressure in white ash. Plant Physiol 68: 121–126

Sperry J 1983 Observations on the structure and function of hydathodes in Blechnum lehmannii ieron. Amer. Fern J. (in press)

Stahl E 1897 Über den Pflanzenschlaf und verwandte Erscheinungen. Bot Ztg 55: 71–108

van Steenis CGGJ 1969 Plant speciation in Malesia, with special reference to the theory of non-adaptive saltatory evolution. Biol J Linn Soc 1: 97–133

Stipes RJ, Campana RJ (eds) 1981 Compendium of elm diseases. Am Phytopathol Soc, 96 pp

Stocker O 1928 Der Wasserhaushalt ägyptischer Wüsten- und Salzpflanzen. Bot Abhandl (Goebel, ed) Vol 2 13: 1–200

Stocker O 1952 Grundriß der Botanik. Springer, Berlin Göttingen Heidelberg

Strugger S 1940 Studien über den Transpirationsstrom im Blatt von Secale cereale und Triticum vulgare. Z Bot 35: 97–113

Sucoff E 1969 Freezing of conifer xylem and the cohesion-tension theory. Physiol Plant 22: 424–431

Suhayda CG, Goodman RN 1981 Infection courts and systemic movement of 32P labeled Erwinia amylovora in apple petioles and stems. Phytopathology 71: 656–660

Sutton JC, Williams PH 1970 Relation of xylem plugging to black rot lesion development in cabbage. Can J Bot 48: 391–401

Swanson RH, Whitfield WA 1981 A numerical analysis of heat pulse velocity theory and practice. J Exp Bot 32: 221–239

Talboys PW 1978 Dysfunction of the water system. In: Horsfall JG, Cowling EB (eds) Plant disease: an advanced treatise Vol 3. Academic Press, New York London, pp 141–162

Thimann KV, Satler SO 1979 Relation between leaf senescence and stomatal closure: senescence in light. Proc Natl Acad Sci USA 76: 2295–2298

Thoday D, Sykes MG 1909 Preliminary observations on the transpiration current in submerged waterplants. Ann Bot 23: 635–637

Thomas DL 1979 Mycoplasmalike bodies associated with lethal declines of palms in Florida. Phytopathol 69: 928–934

Thut HF 1932 The movement of water through some submerged plants. Am J Bot 19: 693–709

Tippett JT, Shigo AL 1981 Barrier zone formation: a mechanism of tree defense against vascular pathogens. IAWA Bull 2: 163–168

Tobiessen P, Rundel PW, Stecker RE 1971 Water potential gradient in a tall Sequoiadendron. Plant Physiol 48: 303–304

Tomlinson PB 1978 Some qualitative and quantitative aspects of New Zealand divaricating shrubs. N Z J Bot 16: 299–309

Tomlinson PB 1982 Anatomy of the monocotyledons. VII. Helobiae (Alismatidae). Clarendon, Oxford 559 pp

Tomlinson PB, Zimmermann MH 1969 Vascular anatomy of monocotyledons with secondary growth – an introduction. J Arnold Arbor 50: 159–179

Traube M 1867 Experimente zur Theorie der Zellenbildung und Endosmose. Arch Anat Physiol Wiss Med 1867: 87–165

Trendelenburg R 1939 Das Holz als Rohstoff. Lehmann, München Berlin

Tyree MT, Cameron SI 1977 A new technique for measuring oscillatory and diurnal changes in leaf thickness. Can J For Res 7: 540–544

Tyree MT, Zimmermann MH 1971 The theory and practice of measuring transport coefficients and sap flow in the xylem of red maple (Acer rubrum). J Exp Bot 22: 1–18

Tyree MT, Caldwell C, Dainty J 1975 The water relations of hemlock (Tsuga canadensis). V. The localization of resistances to bulk water flow. Can J Bot 53: 1078–1084

Tyree MT, Graham MED, Cooper KE, Bazos LJ 1983 The hydraulic architecture of Thuja occidentalis. Can J Bot (in press)

Unger F 1861 Beiträge zur Anatomie und Physiologie der Pflanzen. Sitzungsber math naturwiss Kl Akad Wiss Wien 44 Bd 2: 327–368

Ursprung A 1913 Über die Bedeutung der Kohäsion für das Saftsteigen. Ber Dtsch Bot Ges 31: 401–412

Ursprung A 1915 Über die Kohäsion des Wassers im Farnanulus. Ber Dtsch Bot Ges 33: 153–162

Valverde RA, Fulton JP 1982 Characterization and variability of strains of southern bean mosaic virus. Phytopathology 72: 1265–1268

VanAlfen NK, MacHardy WE 1978 Symptoms and host-pathogen interactions. In: Sinclair WA, Campana RJ (eds) Dutch elm disease: perspectives after 60 years. Cornell Univ Exp Stn Search Agric 8(5): 20–25

Vité JP, Rudinsky JA 1959 The water-conducting system in conifers and their importance to the distribution of trunk-injected chemicals. Contrib Boyce Thompson Inst 20: 27–38

van Vliet GJCM 1976 Radial vessels in rays. IAWA Bull 1976/3: 35–37

van Vliet GJCM 1978 Vestured pits of Combretaceae and allied families. Acta Bot Neerl 27: 273–285

deVries H 1886 Studien over zuigwortels. Maandbl Natuurwet 13:53–68 (Summary in Bot Z 44:788–790)

Waisel Y, Agami M, Shapira Z 1982 Uptake and transport of ^{86}Rb, ^{32}P, ^{36}Cl, and ^{22}Na by four submerged hydrophytes. Aquatic Bot 13: 179–186

West DW, Gaff DF 1976 Xylem cavitation in excised leaves of Malus sylvestris Mill and measurement of leaf water status with the pressure chamber. Planta 129: 15–18

Wieler A 1888 Über den Anteil des sekundären Holzes der dikotylen Gewächse an der Saftleitung und über die Bedeutung der Anastomosen für die Wasserversorgung der transpirierenden Flächen. Jb Wiss Bot 19: 82–137

Wieler A 1893 Das Bluten der Pflanzen. Beitr Biol Pflanz 6: 1–211

Wilson K 1947 Water movements in submerged aquatic plants, with special reference to cut shoots of Ranunculus fluitans. Ann Bot 11: 91–122

Woodhouse RM, Nobel PS 1982 Stipe anatomy, water potentials, and xylem conductances in seven species of ferns (Filicopsida). Am J Bot 69: 135–140

Worrall JJ, Parmeter JR Jr 1982 Formation and properties of wetwood in white fir. Phytopathology 72: 1209–1212

Zajaczkowski S, Wodzicki T, Romberger JA 1983 Auxin waves and plant morphogenesis. In: Scott TK (ed) Functions of hormones in growth and development at levels of organization from the cell up to the whole plant. Encyclopedia of plant physiology New Ser Vol 10. Springer, Berlin Heidelberg New York

Ziegenspeck H 1928 Zur Theorie der Wachstums- und Bewegungserscheinungen bei Pflanzen. Bot Arch 21: 449–647

Ziegler H 1967 Biologische Aspekte der Kernholzbildung. Proc 14th IUFRO Congr, Munich 9: 93–116

Ziegler H 1968 Biologische Aspekte der Kernholzbildung. Holz Roh Werkst 26: 61–68

Zimmermann MH 1960 Longitudinal and tangential movement within the sieve-tube system of white ash (Fraxinus americana L). Beih Z Schweiz Forstver 30: 289–300

Zimmermann MH 1964 Effect of low temperature on ascent of sap in trees. Plant Physiol 39: 568–572

Zimmermann MH 1965 Water movement in stems of tall plants. In: 19th Symposium of the Soc for Exp Biol. The state and movement of water in living organisms. Cambridge Univ Press, pp 151–155

Zimmermann MH 1971 Dicotyledonous wood structure made apparent by sequential sections. Film E 1735 (Film data and summary available as a reprint) Inst wiss Film, Nonnenstieg 72, 34 Göttingen, West Germany

Zimmermann MH 1976 The study of vascular patterns in higher plants. In: Wardlaw IF, Passioura JB (eds) Transport and transfer processes in plants. Academic Press, New York London, pp 221–235

Zimmermann MH 1978a Hydraulic architecture of some diffuse-porous trees. Can J Bot 56: 2286–2295

Zimmermann MH 1978b Structural requirements for optimal water conduction in tree stems. In: Tomlinson PB, Zimmermann MH (eds) Tropical trees as living systems. Cambridge Univ Press, London New York Melbourne, pp 517–532

Zimmermann MH 1979 The discovery of tylose formation by a Viennese lady in 1845. IAWA Bull 1979/ 2-3: 51–56

Zimmermann MH 1982 Functional Anatomy of Angiosperm Trees. In: Baas P (ed) New Perspectives in Wood Anatomy. Nijhoff, The Hague, pp 59–70

Zimmermann MH, Brown CL 1971 Trees. Structure and function. Springer, New York Heidelberg Berlin, 336 pp

Zimmermann MH, Jeje AA 1981 Vessel-length distribution in stems of some American woody plants. Can J Bot 59: 1882–1892

Zimmermann MH, Mattmuller MR 1982a The vascular pattern in the stem of the palm Rhapis excelsa I. The mature stem Film C 1404. (Film data and summary available as a reprint) Inst wiss Film, Nonnenstieg 72, 34 Göttingen, West Germany

Zimmermann MH, Mattmuller MR 1982b The vascular pattern in the stem of the palm Rhapis excelsa II. The growing tip Film D 1418. (Film data and summary available as a reprint) Inst wiss Film, Nonnenstieg 72, 34 Göttingen, West Germany

Zimmermann MH, McDonough J 1978 Dysfunction in the flow of food. In: Horsfall JG, Cowling EB (eds) Plant disease. An advanced treatise Vol 3 Academic Press, New York London, pp 117–140

Zimmermann MH, Milburn JA 1982 Transport and storage of water. In: Lange OL, Nobel PS, Osmond CB, Ziegler H (eds) Physiological plant ecology II. Encyclopedia of Plant Physiology New Ser Vol 12B. Springer, Berlin Heidelberg New York, pp 135–151

Zimmermann MH, Potter D 1982 Vessel-length distribution in branches, stem, and roots of Acer rubrum. IAWA Bull 3: 103–109

Zimmermann MH, Sperry JS 1983 Anatomy of the palm Rhapis excelsa IX. Xylem structure of the leaf insertion. J Arnold Arbor (in press)

Zimmermann MH, Tomlinson PB 1965 Anatomy of the palm Rhapis excelsa. I. Mature vegetative axis. J Arnold Arbor 46: 160–180

Zimmermann MH, Tomlinson PB 1966 Analysis of complex vascular systems in plants: optical shuttle method. Science 152: 72–73

Zimmermann MH, Tomlinson PB 1967 Anatomy of the palm Rhapis excelsa IV. Vascular development in apex of vegetative aerial axis and rhizome. J Arnold Arbor 48: 122–142

Zimmermann MH, Tomlinson PB 1968 Vascular construction and development in the aerial stem of Prionium (Juncaceae). Am J Bot 55: 1100–1109

Zimmermann MH, Tomlinson PB 1969 The vascular system of Dracaena fragrans (Agavaceae) I. Distribution and development of primary strands. J Arnold Arbor 50: 370–383

Zimmermann MH, Tomlinson PB 1970 The vascular system of Dracaena fragrans (Agavaceae) II. Distribution and development of secondary vascular tissue. J Arnold Arbor 51: 478–491

Zimmermann MH, Tomlinson PB 1974 Vascular patterns in palm stems: Variations of the Rhapis principle. J Arnold Arbor 55: 402–424

Zimmermann MH, Tomlinson PB, LeClaire J 1974 Vascular construction and development in the stems of certain Pandanaceae. Bot J Linn Soc 68: 21–41

Zimmermann MH, McCue KF, Sperry JS 1982 Anatomy of the palm Rhapis excelsa VIII. Vessel network and vessel-length distribution in the stem. J Arnold Arbor 63: 83–95

Zweypfenning RCVJ 1978 A hypothesis on the function of vestured pits. IAWA Bull 1978/1: 13–15

Subject Index

Abies
  hydraulic architecture   66–68, 70–72
  wetwood   109
  xylem efficiency   15
Acer
  hydraulic architecture   70, 72, 77, 85
  vessel efficiency   15
  vessel length   12, 13, 34, 35, 88
  vessel structure   91
  water storage   49
Aesculus, effect of girdling   121
Agathis, tracheid length   6
Air ducts in
  aquatic plants   94, 95
  palm stems   98
  wood   50–52
Alpinia, vascular system   25
Aneimia, tensile water   46
Anemone, embolism   115
Apical dominance   67, 82
Aquapot   62
Aquatic angiosperms   92–95

Betula
  flow velocities   38
  hydraulic architecture   71, 72
  vessel diameters   74
  vessel structure   91
  xylem efficiency   15
  xylem pressure   37
Blechnum, water relations   56
Branching, dichotomous (see also hydraulic architecture)   82
Brassica, black rot lesions   112
Bubble manometer   59, 60
Bubbles and
  freezing   80
  pressure   40–47

Calamus, embolism   99, 100
Capillary equation   44, 45
Carinal canals   95
Casparian strip   55–56
Castanea, cavitation   97, 120

Cavitation
  caused by fungi   109–122
  effect on pressure measurements   52, 53
  in aging xylem   96–98
  in fern sporangia   46
  induced by bottle neck   74–80
  recovery from   43, 46
  regulating water flow   54
  relation to pore size   44, 45
Cedrela, vessel network   22–25
Cell length, see tracheid length
Ceratocystis
  direction of movement   111
  effect on *Pinus*   122
  effect on *Ulmus*   118–120
Chrysalidocarpus, water transport   31
Cinematographic analysis   6, 21–23
Click method   40, 43, 44
Cohesion theory   39
Combretum, vestured pits   19

Dendrographs   47
"Designed leaks"   44–47, 54
Dianthus, effect of *Phialophora*   122, 123
Diffuse-porous trees, vessel lengths   11–13
Dioon, vessels   5
Diospyros, heartwood   104
Diurnal stem shrinkage   37, 38, 47, 48
Diurnal water conduction   37, 38, 47, 48
Divaricating shrubs   82
Double-sawcut experiment   34–36
Douglas fir, xylem pressures   60, 61
Dracaena, vascular system   25
Dye movement in xylem   33, 117–119

Earlywood
  flow rate   13
  vessel length   13

Efficiency vs. safety   15, 16
Embolism (see cavitation)
Endodermis in desert plants   55–56
Endothia, causing embolism   120
Equisetum
  carinal canals   95
  spore ejection   46
  tensile water   46
Eucalyptus, pressure gradients   61
Euphorbia, water storage   48

Fagus
  bark disease   121
  vessel network   32
Fern spore ejection   42, 43, 46
Fiber evolution   4
Flooding of leaf parenchyma   56
Flow efficiency vs. safety   14
Fraxinus
  cavitation   52, 53, 97, 115
  flow velocities   38
  narrow-vessel species   65
  vessel network   21, 23
  water storage   52, 53
Freezing of xylem   98–101

Girdling effect on xylem   120, 121
grapevine, pressure gradients   60
gums   103, 104
guttation   37

Hagen-Poiseuille equation   13, 14
Heartwood formation   104–106
"Huber value"   66–68
Hydraulic architecture of
  aquatic plants   67
  conifers   66, 68, 69
  desert plants   67, 82
  dichotomous trees   82
  dicotyledons   70–77

Hydraulic architecture of
    palms   70, 77–81
    Rhynia   68
Hyphae, entry into xylem   110–
    112

Ilex, vessels   12
Impatiens, flow velocities   65
Injection into xylem   33, 123–
    125
Intercellular spaces,
    wettability   56

Land plants, earliest   4
Latewood
    flow rate   13
    vessel length   13
Leaf abscission   81
Leaf insertions   76–80
Leaf-specific conductivity   69–
    76
Leonardo da Vinci, flow
    theory   66, 67

Myriophyllum
    function of roots   92
    vessels   94

Nectria, effect on beech   121
Nelumbo, vessels   94

Palm
    leaf abscission   80, 81
    vascular structure   77–80
Parashorea, vestured pits   19
Pathogen movement   109–112
Peperomia, water storage   48
Perforation plates   89–91
Phaseolus, virus movement
    111
Phialophora, effect on water
    transport   122, 123
Philodendron, water storage   49
Phyosiphon, water storage   48
Picea, water content   96
Pinus
    cambial activity   121
    effect of *Ceratocystis*   122
    KCl effect   69
Pipe model   66
Piratinera, heartwood   104
Pistia, roots   92
Pith, tensions in   46
Pits
    bordered   57–59, 105
    intervessel   7, 10, 16–20, 30
    vestured   19, 20

Populus
    hydraulic architecture   70,
        72, 73, 77
    leaf insertion   77
    path of water   24, 75
    vessel network   21
    wood structure   17
Potamogeton, water transport
    92, 93
Pressure gradients
    and direction of movement
        33
    in plants   56, 59–62, 73
    in *Populus*   73
    problems of measuring   52,
        53, 61
Pressure measurement and
    capillary storage   52, 53
Pressures
    in ferns   56
    negative   37–47
    and pore sizes   56–57
    positive   37
Prionium, vascular system   25
Prosopis, wood structure   57

Quercus
    cavitation   97
    narrow-vessel species   65
    vessel length   11, 15, 35, 36,
        65

Radial transport   83–89
Ranunculus, water transport
    92
Refilling in fall   49
Rhapis
    KCl effect   70
    leaf-trace attachment   78–80
    vascular bundles   25–31, 78–
        80
    vessel lengths   30
    vessels   10, 12, 25–31, 78–80
Rhynia, hydraulic architecture
    68
Ricinus, cavitations in xylem
    43
Ring-porous trees
    vessel length   11
    see also individual species
Robinia
    embolism   101, 115
    tyloses   102, 103
Roots in aquatic plants   92, 93
Roystonea
    leaf abscission   81
    vascular structure   98

Safety
    vs. efficiency   15, 16
    features in dicot wood   34–
        36
Sambucus, tensions in pith   46
Sanchezia, tensions   39
Sealing of xylem   54–59
Segmentation, significance
    80–82
Selaginella
    spore ejection   46
    tensile water   46
Sequoia
    air permeability   50
    tracheid length   87
Sequoiadendron, xylem
    pressures   86
Sharpshooters   111
Sphagnum, water storage   53
Sphenophyllum, tracheid
    length   6
Spiral grain   23–25, 32
Storage of water
    by capillary   48–54
    by elastic expansion   47, 48
    in fibers and tracheids   51–
        54
    in *Sphagnum*   53
Support and water conduction
    4

Tangential spread of water
    33–36
Taxus, flow efficiency   15
Tensile strength
    greatest measured   46
    of water   39–47
Tension limits   44–46
Thermoelectric flow
    measurements   64, 65
Thuja
    hydraulic architecture   72
    KCl effect   69
    transpiration   39
Tracheids
    evolution   4, 5
    in early plants   4
    lengths   4, 6, 86, 87
    network   32
Transpiration
    against vacuum   115–120
    interruption   113–114
Trochodendron, tracheid   5
Tsuga
    bordered pit   18
    hydraulic architecture   70,
        72

KCl effect 69
Tyloses 101–103

Ulmus
  cavitation 97, 117–119
  injection 124
  wavy grain 36
  wetwood 108

Vaccinium, vessel length 12
Valve action of bordered pits
  58, 59
Vasicentric parenchyma 57
Velocities of water movement
  and vessel diameter 75
  in different trees 62–65
Verticillium, effect on water
  transport 122
Vessel
  cross walls 88
  distribution 6, 34, 35
  efficiency 14, 15
  elements length 4, 5
  ends 6, 35, 36
  evolution 4

flow rate in 13
monocotyledonous 25–31
network 7, 21–25
numbers of 16
overlaps 7, 33–34
in palms 10, 25–31, 77–80
ray contact 34
safety 15, 16
significance of width 39
Vessel diameter
  along trunk 74
  at branch insertion 75
  effect on flow 14
Vessel length
  distribution 7–13
  in different plants 30, 34–36
  related to age 89
  within growth ring 89
  within specimen 88
Vessels, solitary 33
Vessel-to-vessel
  contact 29, 33, 34
  permeability 10
  pits 7, 10, 16–20, 30
Vestured pits 19

Vitis
  tyloses 102
  vessel lengths 11
  xylem pressures 37, 60
Vochysia, vestured pits 19

Wall sculptures 89–91
Water conduction
  against vacuum 38–39
  in aquatic pants 92–95
  diurnal 37, 38
  link with mechanical
    support 4
  radial 83–85
  through bordered pits 58
Water storage, see Storage
Wetwood 107–109

Xanthomonas, causing black
  rot 112
Xylem blockage 114–122, see
  also cavitation
Xylem differentiation, pathogen
  effect 123
Xylem vulnerability 121

M. H. Zimmermann, C. L. Brown

Trees

Structure and Function

With a chapter on irreversible thermodynamics of transport phenomena by M. T. Tyree.

(Springer Study Edition)
4th printing. 1980. 134 figures. XII, 336 pages.
ISBN 3-540-07063-X

Contents: Primary Growth. – Secondary Growth. – Growth and Form. – Transport in the Xylem. – Transport in the Phloem.– The Steady State Thermodynamics of Translocation in Plants. – Storage, Mobilization and Circulation of Assimilates.

This second printing of the 1971 edition is devoted largely to those aspects of tree physiology peculiar to tall woody plants. Throughout the book the emphasis is on function as it relates to structure. The authors describe how trees grow, develop, and operate, not how these functions are modified in different types of environment. The approach is functional rather than ecological – how trees work, not how they behave in various habitats.

F. Hallé, R. A. A. Oldeman, P. B. Tomlinson

Tropical Trees
and Forests

An Architectural Analysis

1978. 111 figures, 10 tables. XV, 441 pages
ISBN 3-540-08494-0

Contents: Introduction: What is a Tree? The Botanical World of the Tropics. – Elements of Tree Architecture: The Initiation of the Tree. Apical Meristems and Buds. Extension Growth in Tropical Trees. Phyllotaxis and Shoot Symmetry. Branching: Dynamics. Branch Polymorphism: Long Shoots. Branch Polymorphism: Short Shoots. Abscission. Inflorescence. Raidal Growth: Conifers and Dicotyledons. Radial Growth: Some Variations. Root Systems in Tropical Trees. – Inherited Tree Architecture: The Concept of Architecture and Architectural Tree Models. Illustrated Key to the Architectural Models of Tropical Trees. Descriptions of Architectural Tree Models. Architecture of Lianes. Architecture of Herbs: Miniaturization in Relation to Tree Models. Architecture of Fossil Trees. – Opportunistic Tree Architecture: Reiteration. Energetics. Growth Potential of Forest Trees. A Note on Floristics. – Forests and Vegetation: The Architecture of Forests Plots. Sylvigenesis. – Concluding Remarks. – Glossary. – References. – Index of Plant Names and Their Models. – Subject Index.

Springer-Verlag
Berlin
Heidelberg
New York
Tokyo